Atomistic Simulation of Anistropic Crystal Structures at Nanoscale

Authored by Jia Fu

Published in London, United Kingdom

IntechOpen

Supporting open minds since 2005

Atomistic Simulation of Anistropic Crystal Structures at Nanoscale
http://dx.doi.org/10.5772/intechopen.84596
Authored by Jia Fu

Contributors
Jia Fu

First published in London, United Kingdom, 2019 by IntechOpen
IntechOpen is the global imprint of INTECHOPEN LIMITED, registered in England and Wales, registration number: 11086078, The Shard, 25th floor, 32 London Bridge Street
London, SE19SG – United Kingdom
Printed in Croatia

British Library Cataloguing-in-Publication Data
A catalogue record for this book is available from the British Library

Additional hard copies can be obtained from orders@intechopen.com

Atomistic Simulation of Anistropic Crystal Structures at Nanoscale
Authored by Jia Fu
p. cm.
Print ISBN 978-1-83880-201-1
Online ISBN 978-1-83880-202-8
eBook (PDF) ISBN 978-1-83880-648-4

For EU product safety concerns:
IN TECH d.o.o., Prolaz Marije Krucifikse Kozulić 3, 51000 Rijeka, Croatia, info@intechopen.com or visit our website at intechopen.com.

Meet the author

Dr. Fu is currently an associate professor at Xi'an Shiyou University (China), an executive council in Human Materials and Heat Treatment Association, and the CTO of a large industrial company in Shenzhen. He obtained his PhD degree from the Institut National des Sciences Appliquées de Rennes (France) in 2016, before becoming a visiting scholar for 1 year at the Southern University of Science and Technology. His research interests focus on multiscale modeling and computational material science. He has published more than 50 scientific articles and participated in more than three NSFC projects and one CPER project. He has delivered more than 10 speeches at international and domestic conferences. He serves as the associate editor for RSC Advances and Materials Reviews, among others.

Preface

With the development of the modeling and computer technology, the scale of material has been expanded from macro- to micro-, also from micro- to nanoscale. The development of molecular modeling tools to describe and predict the mechanical properties of structural materials reveals to be of undeniable practical importance. At the nanometer scale, anisotropy is the consequence of differences in the directional arrangements of atomic structure, the force between atoms, such as van der Waals force, typical Coulomb force, and other forces.

Till early twenty-first century, the approximate solution of the Schrödinger equation in quantum mechanics field is almost equivalent to the importance of using Newton's laws widely to solve classical mechanics problems. Density functional theory (DFT) as a first-principles theory and a solid band theory in quantum mechanics has own a great success in linking physical properties and molecular structure, the calculation with exact accuracy but for low computational efficiency for macromolecular structure. Molecular dynamics (MD) has obvious advantages in simulating macromolecular structure, with very high computational efficiency but for much dependency of atomic potential parameters. Besides, MD method is limited into a range of structure at atomic scale level and it cannot effectively solve continuum mechanics problems like practical engineering application using finite element method (FEM). So we look for a modeling tool so as to further widen the application fields of nanoindentation simulation with consideration of atomic forces at nanoscale level. In addition to the tradition of DFT and MD methods that we briefly introduced above, a so-called AFEM modeling methodology at the atomic scale (or nanoscale) is further investigated since the macroscopic properties depend largely on the physicochemical properties of the interatomic bonds.

Nanoscale modeling and mechanical properties by using the density functional theory (DFT), a so-called atomic finite element method (AFEM), and the classical molecular dynamics (MD) method are especially concerned according to the modeling requirement of different crystals structures investigated.

This is a book of a professional and systematic scientific research by Dr. Jia Fu.

This book is committed to model and to calculate mechanical properties (Young's modulus especially) of typical anisotropic crystals structures (cubic crystals: CaO and MgO; hexagonal crystals: CH and calcite; monoclinic crystals: 11 Å tobermorite and gypsum; others: CNT and graphene; monolithic structures of C—S—H (I) (Ca/Si = 0.67) and C—S—H (II) (Ca/Si = 1.67)) by programming with using three methods above. Omitting the description of these structures' importance, elastic moduli are separately calculated by either homogenization or curve fitting of the linear portion of the stress-strain curve by the corresponding numerical simulations. At the end, nanoindentation experiments, SEM, and nanoindentation simulations are carried out to determine elastic moduli of several phases in cement pastes.

As crystal structures at nanoscale always tend to exhibit anisotropy, our research objects of typical cement phase structures depending on crystal types and practical research needs in book are mainly divided into five types: cubic crystals (CaO and MgO), hexagonal crystals (calcite and portlandite), monoclinic crystals (11 Å tobermorite and gypsum), monolithic structures: C—S—H (I) (Ca/Si = 0.67) and C—S—H (II) (Ca/Si = 1.67), and other crystals structures (CNT, graphene). These structures are typical constituents of the microscale Portland hydrated cement paste (HPCP) and their Young's moduli are needed in a 3D multiscale mortar model. Nanoscale modeling helps build the bottom-up multiscale model, by which the accurately independent models of typical structures can be verified by the elastic moduli obtained at atomic scale, then the needed multiscale model can be established. In this book, we mainly focus on modeling and calculating the mechanical properties of these typical anisotropic crystal structures at nanoscale with a high efficiency, then determining homogenized elastic moduli of these structures with a larger scale.

In summary, by identifying relevant parameters using the improved AFEM method and other computational program, it therefore connects nanoscale modeling and continuous pattern of deformation behavior from small scales to larger scales. We believe that our research above will be expanded the existing research and to provide further detailed parameters of our research objects, thus leads to a certain practical significance.

Jia Fu
Xi'an Shiyou University,
Xi'an, China

INSA de Rennes,
Rennes, France

Contents

Chapter 1

Overview of Nanostructure Modeling and Atomistic Simulation Methods

Jia Fu

Abstract

The development of molecular modeling tools to describe and predict the mechanical properties of structural materials reveals an undeniable practical importance. It is now well recognized that such an objective can be achieved through the linking of the structure of materials at the nanoscale or with their performances. At the nanometer scale, anisotropic materials exhibit differences due to the directional arrangements of atomic structure, the force between atoms, such as van der Waals force, typical Coulomb force, and other forces. Nanoscale modeling and mechanical properties by using the density functional theory (DFT), a so-called atomic finite element method (AFEM), and the classical molecular dynamics (MD) method are especially concerned according to the modeling requirement of different crystal structures investigated. Omitting the description of these structures' importance, elastic moduli are separately calculated by either homogenization or curve fitting of the linear portion of the stress-strain curve by the corresponding numerical simulations. This chapter is committed to introduce the modeling and simulation for calculating mechanical properties (Young's modulus especially) of typical anisotropic crystal structures using three methods (DFT, AFEM, and MD) mentioned above. It is therefore asked to connect to the nanoscale modeling and continuous pattern of behavior by identifying the relevant output data at small scales and bringing the necessary information to higher scales.

Keywords: nanoscale, crystal structures, anisotropic elasticity, DFT, AFEM, molecular dynamics simulation, nanoindentation, mechanical properties

1. Introduction

Nanotechnology is considered "high-tech" since it requires deep knowledge of the system considered at the nanoscale in order to intelligently design. Nanomaterials refer to one dimension in the nanometer range (nanosized particles, atomic clusters etc.) or a basic unit at least among a three-dimensional space (nanotubes, ultrathin films, multilayers, superlattices, etc.). Nanoscience and technology turns up in the late 1980; its basic meaning is to understand and transform the nature in the nanometer size range (10^{-10}–10^{-7} m), directly through the manipulation and arrangements of atoms. Material properties mainly include chemical properties, physical properties (density, viscosity, particle size, specific

IntechOpen

heat, thermal conductivity, linear expansion coefficient, etc.), and mechanical properties (elasticity, hardness, strength, plasticity, ductility, toughness, impact resistance, fatigue limit, etc.). At nanoscale, the crystalline material with the smallest unit holds anisotropy, meaning that crystal has different properties in different directions.

With the development of material science and technology, the scale of material has expanded from macro to micro, also from micro to nanoscale. Nanostructure scientific research at nanometer scale includes elaboration, observation, characterization, and analysis. Modeling of miniaturized materials has a lot of new properties, where characteristic space and time scales correspond to typical simulation methods. Modeling and applications of structures at typical space/time scales [1] are shown in **Figure 1**.

From **Figure 1**, for electronic structure, DFT-QHA method is commonly used, while for atomistic structure, Monte Carlo method or molecular dynamics is used. However, real materials are never perfectly isotropic. At nanoscale, the crystal structure reorientations are not a direct result of the applied stress but are a geometrical requirement. Bulk anisotropy due to crystal orientation is therefore induced by plastic strain and is only indirectly related to stress. Recrystallization during annealing usually changes the crystallographic texture but does not cause randomness [2]. When the grain size is reduced to the nanoscale, its Young's modulus and hardness have great changes. The microscopic deformation mechanism is very complex for nanograins, which depends on the shape of the crystal orientation, surface effect, the substrate effect, the grain boundaries affect, etc. In some cases (e.g., composite materials, single crystal), the differences in properties for different directions are so large that one cannot assume isotropic behavior, which leads to another behavior—anisotropic.

Anisotropic behavior is widespread among the variety of materials, and it has a great meaning in practical engineering application in material science field.

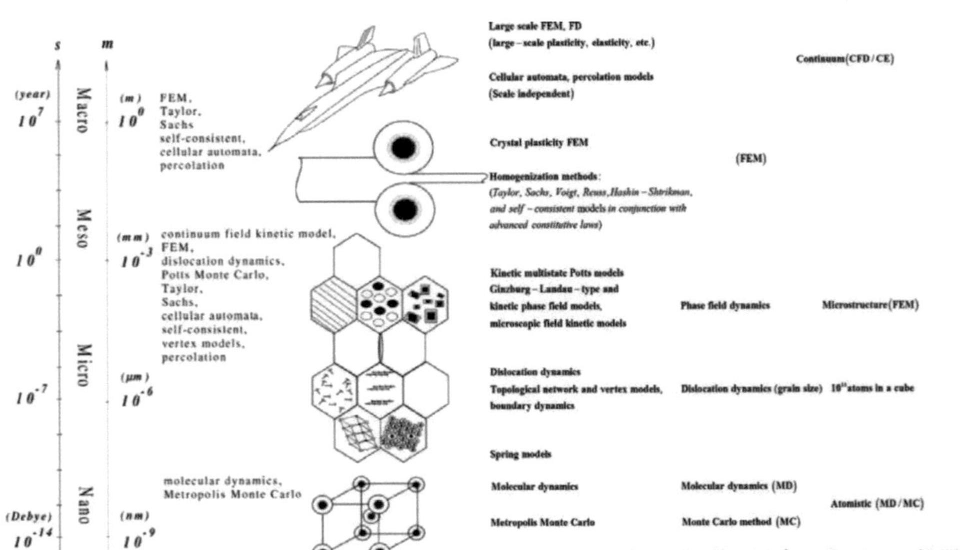

Figure 1.
Modeling and applications of structures at typical space/time scales [1]. (a) Some characteristic space and time scales and (b) characteristic times of typical simulation methods.

Anisotropy is the property of being directionally dependent, as opposed to isotropy, which implies identical properties in all directions. It can be defined as a difference, when measured along different axes, in a material's physical or mechanical properties. From the plastic performance perspective, the anisotropy of plastic deformation of the material shows the performance of texture, ears, and other aspects. Mechanical working produces preferred orientations or crystallographic textures. The primary cause of anisotropy of plastic properties is preferred orientation of grains. In contrast, anisotropy of fracture behavior is largely governed by mechanical alignment of inclusions, voids, and grain boundaries. However, from the elastic performance perspective of crystal structure, the mechanism of elastic anisotropy behavior is still not being recognized, especially for knowing the connection between atomic bonds and mechanical properties of anisotropic crystals structures at nanoscale. Therefore, a comprehensive understanding of the intrinsic nature and the anisotropy behavior at nanoscale has become an urgent task.

Essentially, material properties due to anisotropy are closely related to its internal structure and deformation conditions in the research field of materials science. The microstructure provides a link between processing (how a material is made) and properties (how a material behaves) [3]. It relates with the thermal, electrical, and mechanical properties of a crystal [3]. Alongside the experiment and theory, numerical simulation is a way additional access to the understanding of physical systems. Indeed, it can calculate experimentally measurable quantities and predict properties that are inaccessible in the laboratory or in the model to be validated. In general, both the chemical structure and the microstructure of a material control its properties, of which the chemical structure is relatively fixed and the microstructure depends strongly on how the material is made. However, from mesoscopic level to atomic scale, the drop of material scale makes the contained atoms of nanometer system greatly reduced; besides, macroscopic quasistationary continuous band disappeared, thus showing the energy level separation. The quantum size effect makes physical properties of nanosystem different from other conventional materials, leading to many novel features. At the nanometer scale, anisotropic materials exhibit differences due to these directional arrangements of atomic structure, the force between atoms, such as van der Waals force (caused by hydrogen bonding, ionic bond), typical Coulomb force (caused by charge), and other forces caused by covalent bond, atomic distortion, defects, hydrophilic, etc. Numerical simulation of anisotropic plastic behavior has been investigated by Yonggang [4] in Harvard University, while the numerical simulation and elastic properties of crystal structure at nanoscale have not been investigated systematically.

Although material performance is not the same in different directions in crystal, however, it has a strict symmetry. Anisotropic behavior of crystalline/structure can be reflected in the constitutive relation, where the elastic coefficient matrix corresponding to crystal system is certain. For different crystal/system, the number of cubic crystal has the least number of 3, hexagonal crystal 6, monoclinic crystal 13, and triclinic crystal with the most number of 21. Elasticity coefficient has important practical significance on the scientific basis and engineering applications of the material. Different crystal systems can be characterized exclusively by their symmetries. Each crystal has a certain level of crystal symmetry with corresponding different number of independent elastic coefficients. By obtaining elastic coefficient matrix, homogenized elastic properties (bulk, shear, and Young's moduli) of polycrystals at larger scale can be determined, using the Voigt-Reuss-Hill estimation, for example. Because of complex issues mentioned above, for different anisotropic structure, selecting the appropriate modeling approach is particularly critical.

2. Nanostructural modeling and simulation methods at nanometer scale

With the development of the modeling and computer technology, the scale of material has expanded from macro to micro, also from micro to nanoscale. Nanomechanics aims to study fundamental mechanical properties of material structure at the nanoscale. Nanoscale modeling methods mainly include quantum molecular (QM), Monte Carlo (MC), molecular (structural) mechanics (MM) [5], molecular dynamics (MD) [6], nonlocal continuum theory [7], the ab initio calculation [8], tight-binding molecular dynamics (TBMD), and the density functional theory (DFT) [9]. Typical application, different space scale, the corresponding time scales, and simulation methods in computational materials science fields have been discussed by Raabe [1, 10]. Molecular simulation is also known as molecular modeling, which refers to theoretical methods and computational techniques to model or imitate the behavior of molecular. The model size at different scales has gradually decreased to continuum and mesoscopic level (material model diameter greater than 10^{-4} m), the mesolevel (about 10^{-6}–10^{-4} m), the microlevel (about 10^{-7}–10^{-6} m), and atomic/nanolevel (about 10^{-10}–10^{-7} m). **Figure 2** shows typical scales of material computational mechanics.

From **Figure 2**, scales in computational mechanics field can be divided into four scales: the macro (macroscale), mesoscopic (mesoscale), microscopic (microscale), and nano (nanoscale). Correspondingly, the continuum mechanics is commonly used to solve the macroscale structures, where finite element method can be used. For molecular dynamics for microscale, mesoscale, and nanoscale structures, for example, the discrete model can be an option at mesoscale. For quantum mechanics for both nanoscale and atomic scale structure, for example, density functional theory can be an effective option.

During the last decade, nanomechanics has emerged on the crossroads of classical mechanics, solid-state physics, statistical mechanics, materials science, and quantum chemistry. Numerical methods represent the most versatile computational method for the various engineering disciplines, and the scale of material modeling is gradually transited from bulk scale to nanoscale in **Figure 3**.

From **Figure 3**, the simulation approach for the structure in each domain is not all the same. The internal distance of each structure from quantum scale (nanoscale and atomic scale) to bulk scale covers the range of nanometer to micrometer and the time scale within the range of femtosecond to seconds.

Correspondingly, numerical modeling in computational material mechanics includes several analysis techniques such as ab initio methods, molecular dynamics, finite elements, boundary elements, distinct elements, and other numerical approaches that depend on the material/structure at different scales. Relationship between modeling methods and time/length scale is shown in **Table 1**.

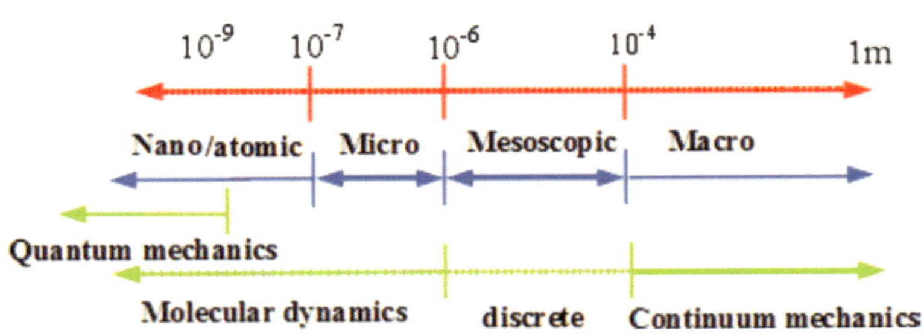

Figure 2.
Scales of material computational mechanics.

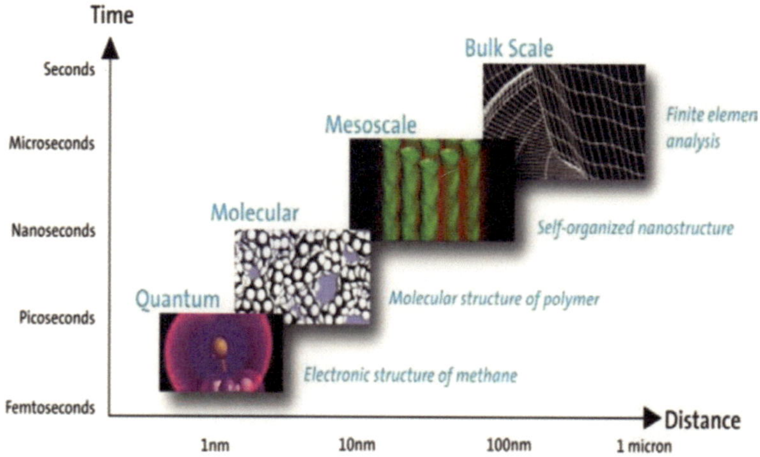

Figure 3.
Length/time scales using various methods in computational materials mechanics.

Table 1 shows typical space, time scales, and simulation methods. These methods can separately link the relationship between modeling methods and real structure size. The fundamental characteristic of numerical methods is that a structure is discretized into small elements. Then the corresponding models that describe the individual elements and their interactions are constructed. The scale of material has expanded from macro to micro, also from micro to nanoscale.

2.1 Classical quantum mechanics and molecular dynamics methods

2.1.1 Brief introduction of quantum mechanics (QM) method

In the early 1900, the basics of quantum mechanics were gradually proposed by Schrödinger and others to study the quantum phenomena when the dimension of the system is in the size of atoms and molecules. Schrödinger equation [11] is proposed and it plays the same role as Newton's equation in classical systems but exactly describes the wave function of a system that evolves with time. Then, some approximations [12], such as the first ab initio Hartree-Fock (HF) calculations on diatomic molecules, were carried out at the Massachusetts Institute of Technology in 1956 [13]; thus, the numerical resolution of the time-independent Schrödinger equation was calculated. These traditional methods are briefly introduced as follows.

First-principles calculations are also called ab initio method, which refers five fundamental constants of physics—mass of the electron, the electron charge, Planck's constant, the speed of light, and Boltzmann constant. The calculation is to compute the system integrating of all the electronics solving the Schrodinger

Modeling methods	Length scales (m)	Time scales (s)
Quantum mechanics method	$< 10^{-8}$	10^{-15}–10^{-12}
Molecular dynamics method	10^{-8}–10^{-6}	10^{-12}–10^{-9}
Micromechanics methods	10^{-6}–10^{-4}	10^{-9}–10^{-3}
Continuum method	$> 10^{-3}$	$> 10^{-3}$

Table 1.
Relationship between modeling methods and time/length scale.

equation without considering any empirical parameters and thus can reasonably predict the status and nature of microscopic systems. Besides, it is a strong theoretical approach developed from quantum mechanics and quantum chemistry, mainly to study hundreds of atoms or periodic crystal material. First-principles calculations can be used not only to explain the physical nature of the experimental phenomena, relationship between microscopic electronic structure and the physical parameters, but also to design new materials and to theoretically predict many physical properties.

Density functional theory is a kind of first-principles molecular dynamics calculations [14], which is also called ab initio molecular dynamics. First-principles molecular dynamics make up shortcomings that the classical molecular dynamics need to know the atomic interaction potential. The atomic interaction potential is solved by DFT [15] and then substituted into the atomic motion equations to solve the trajectory of the particles of the system. The main difference between first-principles molecular dynamics and classical molecular dynamics is to calculate the atomic interaction force [14, 15]. Advantages of first-principles molecular dynamics are as follows: (1) quantum mechanics is used to describe the interaction between electrons and atomic structure and (2) it is able to calculate the effect of nonharmonic crystal thermal vibrations. The downside of DFT is that it is time-consuming [16].

2.1.2 Brief introduction of molecular dynamics (MD) method

The molecular dynamics (MD) simulation to describe the atomic behavior by classical models and equations was reported in 1956 [17]. Molecular dynamics is to calculate a set of molecular orbital phase space. The MD method uses a microclassic calculation, assuming that atomic motion can be described by Newton's motion equations, which means that atomic motion is associated with a particular track. This assumption is feasible and available only when the movement of quantum effects can be ignored and the adiabatic approximation is strictly satisfied. Molecular dynamics is a deterministic method that offers the possibility of a microscopic description of a physical system in consideration of all the interactions involved. The main advantage of this method is that it gives the information on the evolution of the system over time by numerically solving Hamilton equations of motion, Lagrange, or Newton.

Before showing how organized a simulation by molecular dynamics, it is important to define some necessary terms (NVE, NVT, NPH, and NPT ensembles) in such a simulation. The MD method is a deterministic simulation technique for evolving systems to equilibrium by solving Newton's laws numerically. Molecular dynamics method can be used to study the molecule substances with lots of atoms (even up to millions atoms). The implemented potential was tested to reproduce the basic physical parameters of bulk, defects, and surface. The bulk physical parameters, such as the cohesive energy, lattice constant, and bulk modulus, were obtained on the supercell structure with a certain boundary condition, which were equilibrated by MD method with the set of a time step.

2.2 The AFEM methodology based on molecular mechanics method

Molecular mechanics method aims to study fundamental mechanical properties at the atomic scale. It has emerged on the crossroads of classical mechanics, solid-state physics, statistical mechanics, materials science, and quantum chemistry. Length/time scales using various methods in computational materials mechanics are shown in **Figure 5**.

From **Figure 4**, we can see that there is a gap between time record and space record. The QM method, such as the density functional theory (DFT) as a first-principles theory and a solid band theory in quantum mechanics, has own a great success in linking physical properties and molecular structure, the calculation with exact accuracy but for low computational efficiency for macromolecular structure. Molecular dynamics (MD) has obvious advantages in simulating macromolecular structure, with very high computational efficiency but for much dependency of atomic potential parameters. Besides, MD method is limited into a range of structure at atomic scale level, and it cannot effectively solve continuum mechanics problems like practical engineering application using finite element method (FEM). In fact, the physicist Garrett Lisi in 2007 has proposed "a theory of everything" and has proposed an E_8 model [18]. Garrett Lisi believes that all known physics particles and elementary various forces can be accommodated at the E_8 mode where the plurality of interactions between particles is naturally appeared.

So how to establish a connection between quantum-mechanical (QM), molecular dynamics (MD), and continuum mechanics, so as to predict or calculate macroscopic properties at the atomic scale based on some kinds of continuum methods?

These possible modeling methods cause great interest and concern of the scholars. Quantum-mechanical (QM) or ab initio methods mainly aim to solve electronic Schrodinger equation for atoms and molecules, as well as various approximations of that solution. Electronic structure of each atom is accounted for. Major methods include density functional theory, Hartree-Fock, and tight binding. QM methods are very accurate but extremely expensive. Otherwise, without consideration of the complex electronic structure of each atom, the whole atom calculation methods can be realized by modeling the structure/system just as a type of ball (soft sphere), using a series of empirical interatomic potentials. These major methods (molecular mechanics, molecular dynamics, Monte Carlo) are less accurate but relatively inexpensive. Although MD method can simulate the microscopic deformation in a certain degree, some effects due to size are too small but still cannot be obtained because of computational limitations of molecular dynamics simulations, and continuum mechanics finite element method cannot effectively simulate the microstructure deformation behavior. Because of these deficiencies of molecular dynamics and continuum methods, multiscale method came into being,

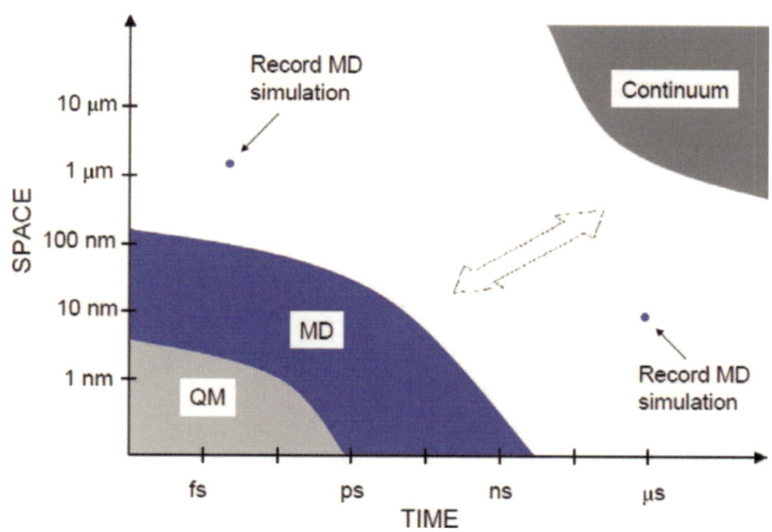

Figure 4.
Length/time scales using various methods in computational materials mechanics.

which can be used to simulate in larger sizes without ignoring the deformation at smaller scale.

In addition to the traditional simulation methods that we briefly introduced above, a so-called atomic finite element method (AFEM) [19] at the atomic scale (or nanoscale) has attracted our great interest. In fact, models of crystal structures at the nanometer scale often tend to exhibit anisotropy. Modeling of anisotropic crystal structure needs to include the atomic and force field information, which leads to anisotropy due to the consequence of differences in the directional arrangements of atomic structure. The force between atoms, such as van der Waals force, typical Coulomb force, and others, should be considered.

In 2011, an atomic finite element (AFEM) as a bridge between molecular dynamics and continuum mechanics is proposed, which makes the classical finite element method (FEM) to model at nanoscale.

Considering Coulomb interaction, the total steric potential energy U by Rappe is as [20]:

$$U = \sum U_r + \sum U_\theta + \sum U_\phi + \sum U_\omega + \sum U_{vdw} + \sum U_{coul} \quad (1)$$

where U_r is bond stretching, U_θ is bond angle bending, U_ϕ is dihedral angle torsion, U_ω is improper torsion, U_{vdw} is nonbonded van der Waals interaction, and U_{coul} is Coulomb (electrostatic) interaction.

At this scale, the interatomic forces are considered binding van der Waals, electrostatic force, and the covalent chemical bond. However, its application is limited as well, including the simple binary alloy and covalent bonds [19]. The application of AFEM in the field of ionic crystal is blank. Description of main force field of molecular mechanics is shown in **Figure 5**.

From **Figure 5**, covalent bonds can be characterized to be equivalent beam by the molecular structural mechanics, and nonlinear spring element can be used to characterize the coupled field of van der Waals force and Coulomb force between atoms or charged ions by using Lennard-Jones potential. For AFEM modeling, the equivalent beam's definition of other covalent bond can be approximated solved by a linear bond energy ratio according to C—C bond energy. This book tries to approach this goal, elastic constants under various anisotropic crystal directions and homogenized elastic moduli of polycrystal become the research priorities of subject.

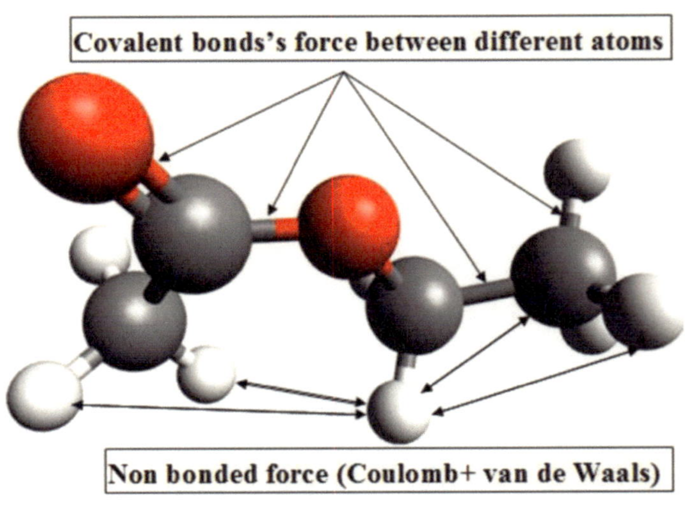

Figure 5.
Description of main force field of molecular mechanics.

In addition, the conventional DFT methods and MD methods are also used to compare mechanical properties and the simulation results of research objects.

As a structural mechanics approach [5] has successfully modeled the frame-like structure of C—C covalent bonds by equivalent beam elements and an AFEM methodology has solved the interatomic bonds modeled by nonlinear spring elements, modeling of nanostructure is realized by an equivalent beam, and nonlinear spring elements become possible. Such modeling tool may be used to further widen the application fields of nanoindentation simulation with consideration of atomic forces at nanoscale level.

3. Research background on the multiscale modeling of concrete platform

Concrete is the world's most widely used man-made material with the history of more than 160 years. It is a complex mixture of numerous compounds, which has been used in the application in the antiseismic building, civil engineering, construction, marine engineering projects, etc. [21] Our research background is based on some typical crystal structure of cement pastes in the reinforcing concrete, which accounts for the platform MuMoCC (Multi-Scale Modeling of Computational Concrete).

3.1 Modeling scales of concrete and cement hydration products

Concrete is a highly heterogeneous composite construction material whose microstructure contains randomly dispersed features. Concrete is primarily composed of cement paste, sand, coarse aggregate, and admixtures. Each of these ingredients is heterogeneous themselves at a certain level and has different stiffness and strength. Therefore, the heterogeneity of concrete exists in a variety of length scales.

3.1.1 Different scales of the concrete modeling

All the types of cement pastes, in general, are truly complex materials, porous, multicomponent, and multiscalar. Moreover, as the hydration products have a very dissimilar structure and sizes, the cement paste itself has different hierarchical levels of organization at different scales. **Figure 6** shows schematic representation of top-down five scale levels model for normal concrete [22, 23], which gives level classification and typical structures of cement-based composite material.

From **Figure 6**, the typical structures can be divided into five levels, from the scale of pavement (10^{-1} m) down to the C—S—H solid phase (10^{-10} m) [22]. Each length scale is a random composite, and the description of the cement paste structure will be presented for each scale. A multiscale level model has been defined based on morphology and length scale of the structural elements [24]. The most usual classification divides the multiscale structure into four size ranges: macroscale, mesoscale, microscale, and nanoscale.

3.1.2 Cement hydration process and typical hydrated structures

Hydration refers to the chemical reactions between cement and water; the hydration process causes cement paste to first set and then harden. Meanwhile, hydration products—some new solid phases—are formed by hydration. Cement hydration refers to a lot of complex mixtures of numerous compounds. The major

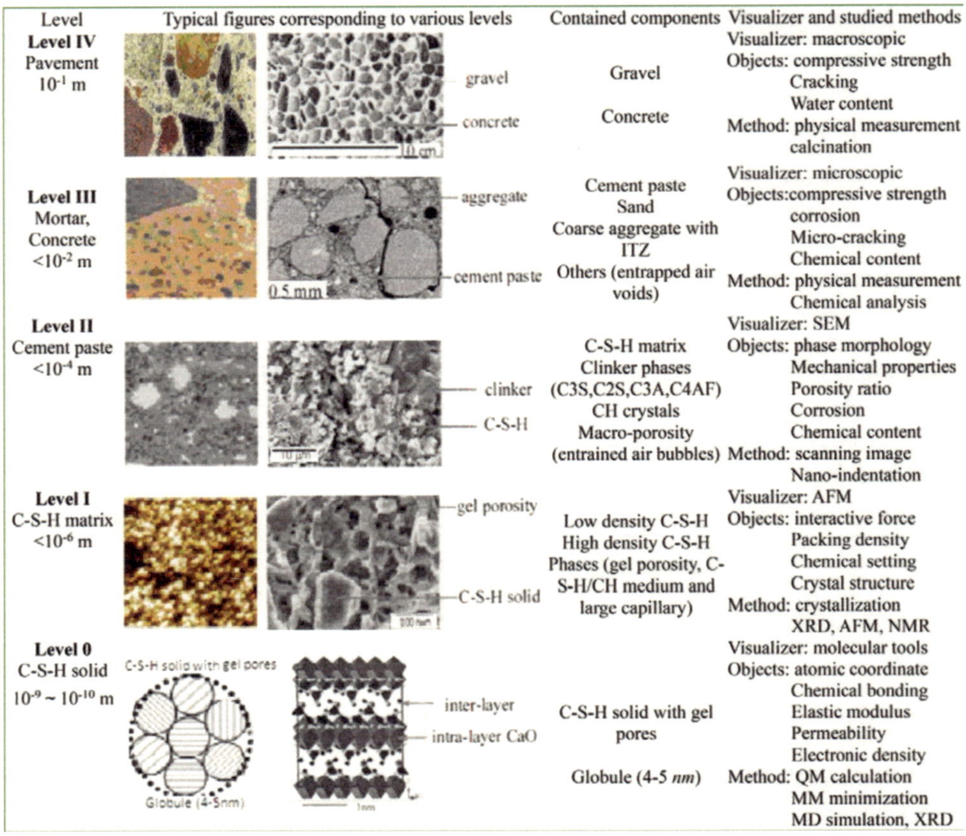

Figure 6.
Level classification and microstructures of cement-based composite material [22, 23].

compounds found in cement are tricalcium silicate (C_3S), dicalcium silicate (C_2S), tricalcium aluminate (C_3A), tetracalcium aluminoferrite (C_4AF), and gypsum. Some tricalcium silicate produces C—S—H gel and calcium hydroxide (CH) is formed at the early stage of hydration, while dicalcium silicate produces C—S—H gel and CH is formed at a later stage in the hydration process.

Figure 7 summarizes the order of the different hydrates and their relative quantities [25].

Figure 7 shows the degree of hydration, particularly at later times. Dashed curve of the porosity decreases in proportion with the rate of hydration of the cement paste, due to the progressive filling empty initial intergranular by various hydrates. The C—S—H gel is the most important product of hydration. Furthermore, it is found that after a few hours, ettringite increases, reaches a maximum, and then decreases in favor of monosulfoaluminate, calcium aluminates, and hydrated aluminates $C_4(A,F)H_{13}$. At the end of stage (after 28 days), the initial cement has hydrated, and the paste has undergone final set.

Typical microstructures and scanning electron microscope (SEM) of cement hydration products by Regourd [26] in 1975 is shown in **Figure 8**.

In **Figure 8**, we can see the SEM image of C—S—H microstructures that are nanocrystalline compounds of nanometric particle aggregated with each other. Among the SEM of a clinker grain, the well-defined side and the pseudohexagonal contour of the upper face of the crystal of C_3S are noted. Aluminates are as a "stick" structure between the crystals C_3S calcium silicates and/or C_2S. Alkali sulfates are condensed on the surfaces of the silicate crystals during solidification process. The

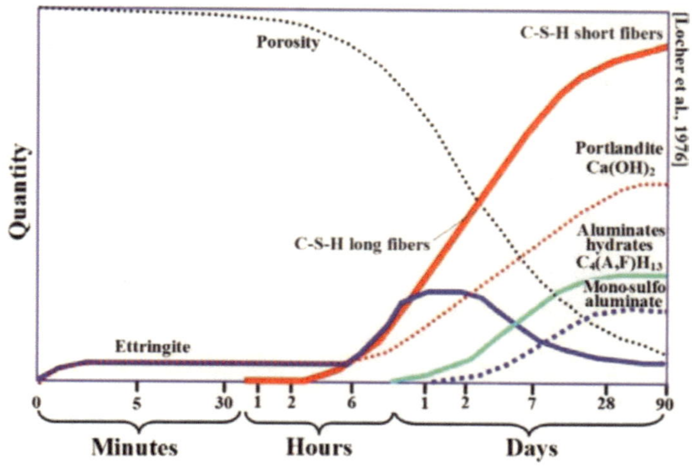

Figure 7.
Crystals of silicates during solidification process [25].

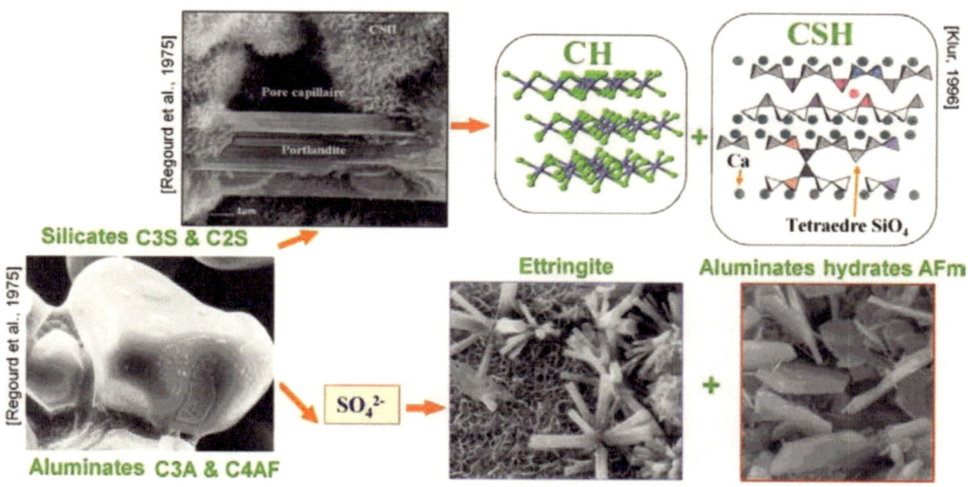

Figure 8.
Typical microstructures of cement hydration products [26].

hydration of C_2S and C_3S forms calcium hydrosilicate (C—S—H), which is accompanied with calcium hydroxide (CH). The continuous and relatively rapid deposition of hydration products (primarily C—S—H gel and CH) into the capillary porosity occurs, which is the space originally occupied by the mix water, leading to a large decrease in the total pore volume and a concurrent increase in strength [26].

C—S—H is responsible for most of the cement paste properties, which is amorphous and heterogeneous in composition. It is well known that the C—S—H gel can be classified as inner product and outer product, of which the inner product of C—S—H gel grows inward the boundaries of the clinker grains, occupying the place of the anhydrous phases, while the outer product of C—S—H gel is formed outward the boundaries of the clinker grains, in the water filled space [27]. In general, the inner product has a more compact and more amorphous structure, while the outer product has been found to form bundles of fibers radiating from the cement grains [28]. The composition of both products seems to be similar, although some authors report slight differences in the Ca/Si ratio, which is higher for the inner product [29].

3.2 Background on reinforcing concrete and MuMoCC platform

3.2.1 Research background on concrete reinforced by nanotubes

Nowadays, the use of self-compacting concrete (SCC) is progressively changing the method of concrete placement on building sites [30]. Concrete is a heterogeneous material composed of a mixture of sand, aggregates embedded in a hardened cement paste. Cement is the binder component of concrete, the glue that holds the filler together to create a uniform, strong material. The cement notation uses letters that symbolize the main components expressed as oxides: CaO (*C*), SiO_2 (*S*), Al_2O_3 (*A*), Fe_2O_3 (*F*), H_2O (*H*), SO_3 (*S*), etc., forming crystal morphology. As a reinforcing material in cement matrix, carbon nanotubes have high stability and significantly improved its mechanical properties, impact resistance, antistatic properties, and wear resistance properties. Therefore, the development of a cement matrix of carbon nanotubes becomes scholars' interest. In 2012, the conception that "cement nanotubes as a natural means for reinforcing concrete [31]" is proposed, shown in **Figure 9**.

From **Figure 9**, it has been concluded that nanotubes make calcium silicate hydrates an ideal cement-reinforcement paste, of which the stress deformation of portlandite nanotubes increases almost linearly up to a strain of 27% [31]. Mechanical properties of cement matrix reinforced by carbon nanotubes are super excellent; it is not only resistance to fatigue and creep but also dimensionally stable. Carbon nanotubes can be formed by rolling a cylindrical graphene and thus can be used as a reinforcing material of brittle materials such as cement.

3.2.2 Research background about MuMoCC platform

Modeling of nanostructure and atomic/ionic crystal and their elastic properties has become a hot field for many scholars [32–41]. Another research background is that a hierarchical multiscale modeling of the behavior of cement-based materials and the multiscale modeling of computational concrete (MuMoCC) platform [42, 43] have been proposed; thus, behavior of hydrates (like C—S—H, CH, etc.) and admixtures (like calcite, CNT, etc.) are needed to be investigated in detail. Elastic properties (Young's modulus, Poisson ratio, etc.), tensile strength, and postfailure tensile stress displacement are thus the unique data that are transferred from cement paste to mortar [43]. Besides, the elastic displacements are found in every pixel, and the average strain and stress is computed and averaged over the entire microstructure to give the effective elastic properties. Input data of these solid phases at the nanoscale are needed. Different scales in structural modeling of concrete using HYMOSTRUC model of multiscale structures by UT Delft [44] and MuMoCC platform [42, 43] are in **Figure 10**.

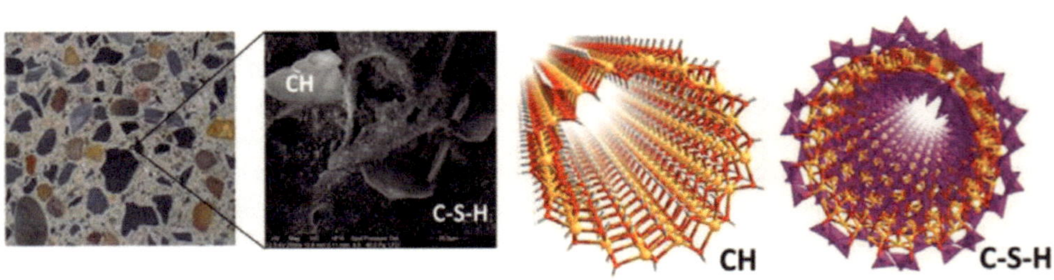

Figure 9.
Main cement paste types and nanotube reinforcing concrete expectation [31].

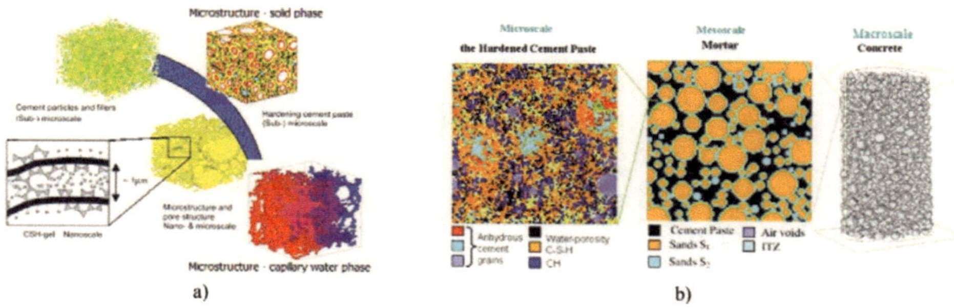

Figure 10.
Different scales in structural modeling of concrete. (a) HYMOSTRUC model of multiscale structures by UT Delft [44] and (b) the representative MuMoCC platform [42, 43].

The methodology has been successfully applied to determine mechanical as well as transport properties of cement-based materials, and also the evolution of these properties according to time, and their modification due to degradation phenomena such as leaching.

The prediction of the mechanical effective properties has been an active research area for the few decades. Numerical models seem to be a well-suited approach to describe the behavior of semianalytical material models, because there is no restriction on the geometry, on the material properties, on the number of phases in the composite, and on the size. Except for the experimental studies, either micro- or macromechanical methods are used to obtain the overall properties of material phases. In the macromechanical approach, the heterogeneous structure of the composite is replaced by a homogeneous medium with anisotropic properties. Micromechanical method provides overall behavior of the matrix-inclusion composites from known properties of their constituents (inclusion and matrix) through an analysis of representative volume element or a unit cell model. The advantage is not only the global properties of the composites but also various mechanisms such as damage initiation and propagation, through the analysis.

4. Objectives and characteristics of typical anisotropic crystals at nanoscale

Under the research background of the conception that "cement nanotubes as a natural means for reinforcing concrete" and the platform construction of MuMoCC (multiscale modeling of computational concrete), 10 kinds of crystals structures are, respectively, studied so as to achieve the calculated values of their mechanical properties (Young's modulus in particular) and to perfect the crystal elastic theory of typical crystal structures simultaneously.

4.1 Objectives of typical anisotropic crystal structures at nanoscale

4.1.1 Nanoscale modeling on concrete reinforced by nanotubes

In 2016, Eftekhari and Mohammadi [45] have investigated the molecular structure of CNT-reinforced C—S—H phase, the C—S—H model is 3.925 nm × 3.626 nm × 4.768 nm, shown in **Figure 11**.

From **Figure 11**, the enhanced mechanical properties of C—S—H reinforced by embedding CNT in its molecular structure are simulated by MD method. It is indicated that the tensile strength of CNT-reinforced C—S—H is substantially

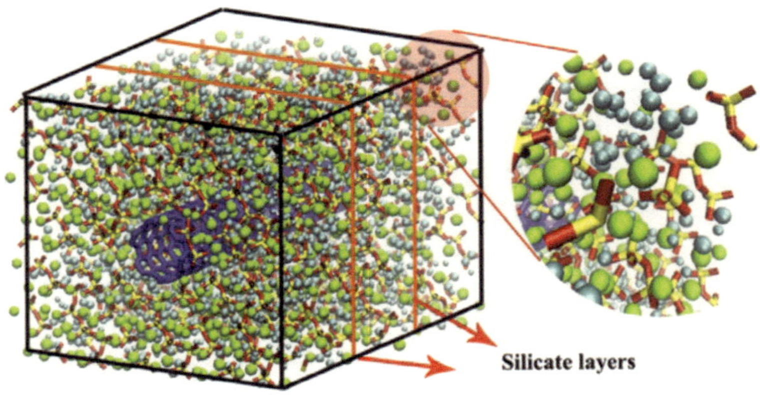

Figure 11.
The molecular structure of CNT-reinforced C—S—H phase [45]. (Atom types: red, oxygen; cyan, hydrogen of water molecules; green, calcium; yellow, silicon; blue, CNT) (diameter of 0.813 nm with length of 4.0 nm).

enhanced along the direction of CNT as compared to the pure C—S—H. In addition, CNTs can severely intensify the "transversely isotropic" response of the CNT-reinforced C—S—H, which can be used for multiscale simulation of crack bridging at macroscale specimen of CNT-reinforced cement.

In all, from a cementitious composite [46] reinforced with carbon CNT and/or polypropylene microfibers, a dense C—S—H formation that appears to be tightly bonded to the SWCNT, producing reinforcing behavior [47]. In fact, nanotubes and graphene as their superior mechanical properties have been more and more confirmed to be the reinforcing materials to enhance the toughness of brittle materials [48].

4.1.2 Objectives of typical anisotropic crystal structures at nanoscale

The use of CNT as a reinforcing material is intended to move the reinforcing behavior from the macroscopic to the nanoscopic level, which has the advantage of extremely high strengths, Young's modulus, elastic behavior, and advantageous thermal properties. Crystal elastoplastic behavior has been the long concern for scholars, with the introduction of nanoscale conception; the performances of anisotropic crystal or polycrystalline become very interesting. The investigation of the crystal elastic simulation has become an urgent issue to be resolved. Combining the introduction mentioned above and the actual research needs, Young's modulus of typical crystal structures is especially concerned, and models of these considered structures are shown in **Figure 12**.

From **Figure 12**, depending on crystal types of typical structures and practical research needs, these crystals structures are divided into cubic crystals (CaO and MgO), hexagonal crystals (calcite and portlandite), monoclinic crystals (11 Å tobermorite and gypsum), monolithic disordered structures (C—S—H (I) with Ca/Si of 0.67 and C—S—H (II) with Ca/Si of 1.67), and other crystals structures (CNT, graphene).

This book mainly includes three types of the cubic, hexagonal, and monoclinic crystals, as well as the other crystal structures considered. Abandoning the lengthy discussion of the important description in these typical structures, we mainly focus on modeling and calculating the mechanical properties of these typical anisotropic crystal structures at nanoscale with a high efficiency and then determining

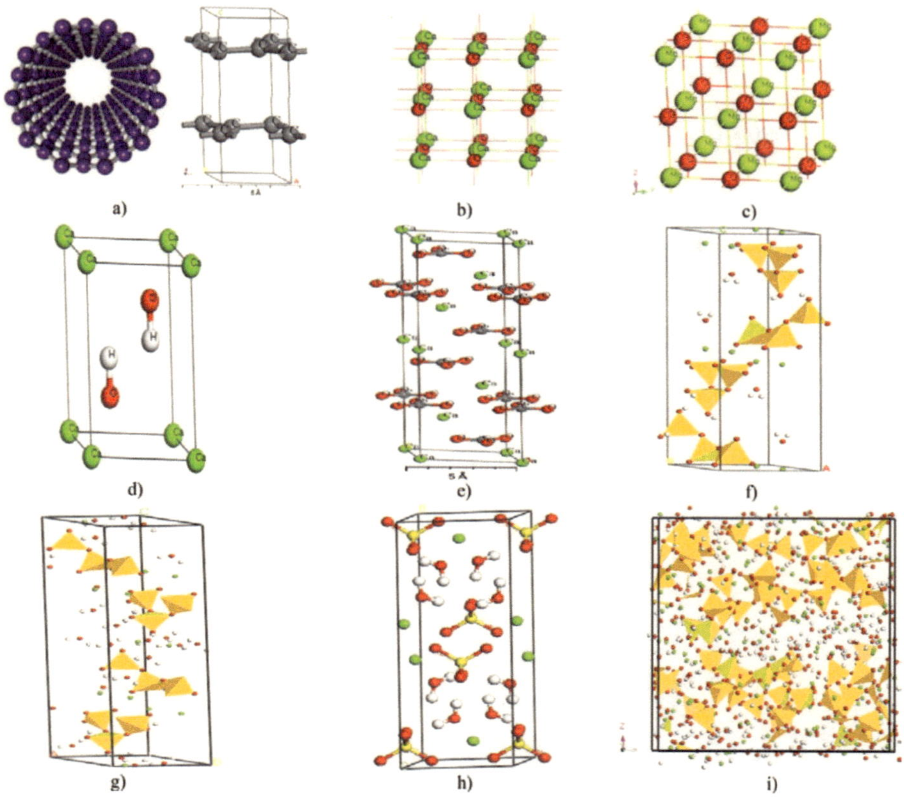

Figure 12.
Crystal modeling objects and corresponding real structures at nanoscale. (a) CNT and hexagonal graphene. (b) Cubic CaO structure. (c) Cubic MgO structure. (d) Hexagonal CH structure. (e) Hexagonal calcite structure. (f) Monoclinic 11 Å tobermorite. (g) Monoclinic 14 Å tobermorite. (h) Monoclinic gypsum structure. (i) Amorphous disordered C—S—H structure.

		Crystal type	Number of C_{ij}	Independent elastic coefficient
		Isotropic	2	c_{11}, c_{12}
a = b = c	α = β = γ = 90°	Cubic	3	c_{11}, c_{12}, c_{44}
a = b ≠ c	α = β = 90°, γ = 120°	Hexagonal	5	c_{11}, c_{12}, c_{13}, c_{33}, c_{44}
a = b ≠ c	α = β = γ = 90°	Tetragonal	6	c_{11}, c_{12}, c_{13}, c_{33}, c_{44}, c_{66}
a ≠ b ≠ c	α = β = 90°, γ = 120	Trigonal	6	c_{11}, c_{12}, c_{13}, c_{33}, c_{14}, c_{44}
a ≠ b ≠ c	α = β = γ = 90°	Orthorhombic	9	c_{11}, c_{22}, c_{33}, c_{12}, c_{13}, c_{23}, c_{44}, c_{55}, c_{66}
a ≠ b ≠ c	α = γ = 90°, β ≠ 90°	Monoclinic	13	c_{11}, c_{22}, c_{33}, c_{12}, c_{13}, c_{23}, c_{44}, c_{55}, c_{66}, c_{16}, c_{26}, c_{36}, c_{45}
a ≠ b ≠ c	α ≠ γ ≠ β ≠ 90°	Triclinic	21	c_{11}, c_{12}, c_{13}, c_{14}, c_{15}, c_{16}, c_{22}, c_{23}, c_{24}, c_{25}, c_{26}, c_{33}, c_{34}, c_{35}, c_{36}, c_{44}, c_{45}, c_{46}, c_{55}, c_{56}, c_{66}

Table 2.
Independent elastic constants for seven kinds of crystal symmetry.

homogenized elastic moduli of these structures with a larger scale. The numerical simulation to determine elastic modulus of these typical structures is a meaningful research.

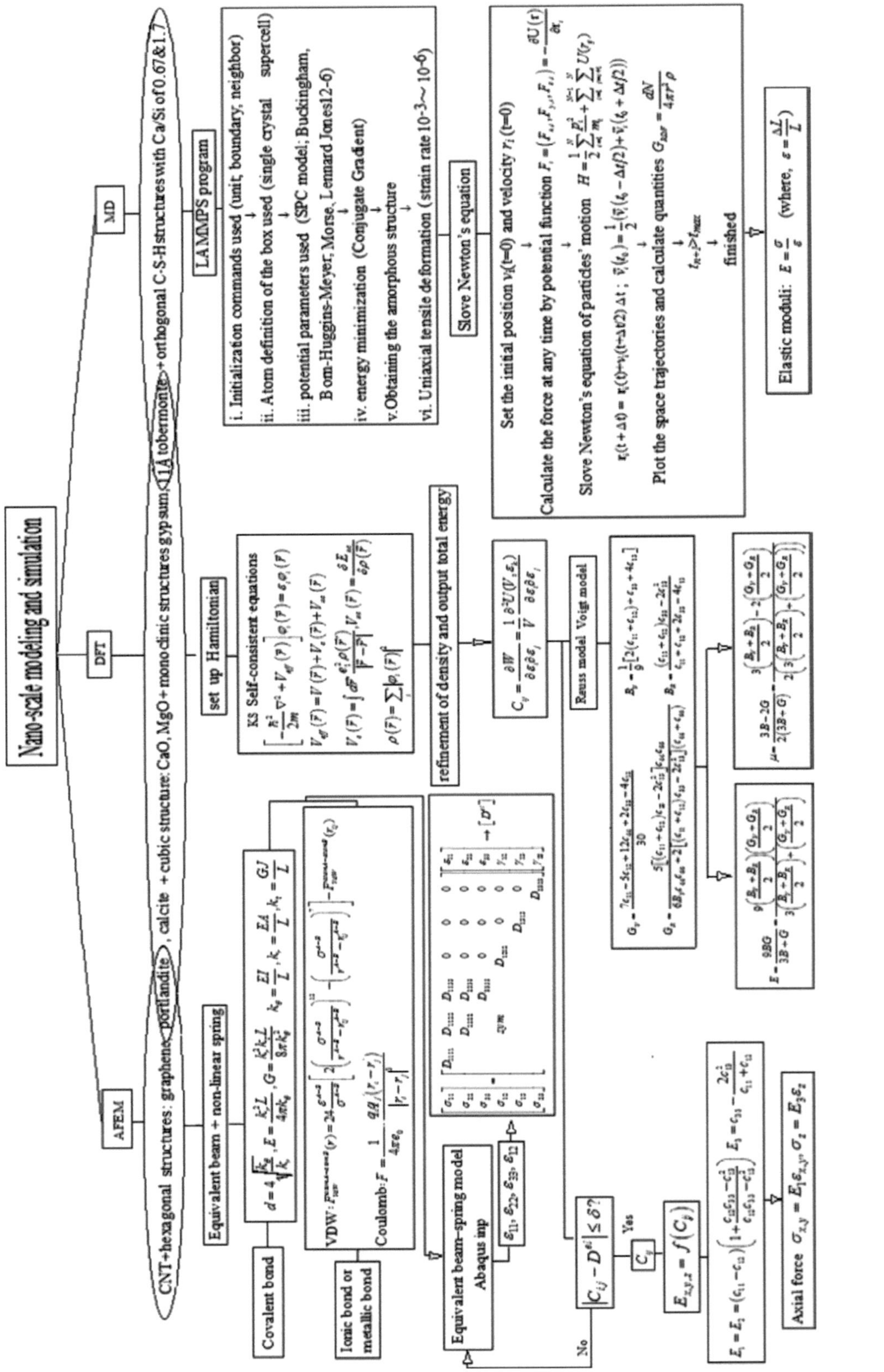

Figure 13.
Elastic moduli determination based on AFEM, DFT, and MD.

4.2 Determination of crystal-independent elastic constant coefficients

In matrix format, the stress-strain relation shows the 36 (6 × 6) independent components of stiffness. Crystal elastic anisotropy can be expressed by a fourth-order tensor matrix of strain-stress relation; Hooke's law thus can be written as a general symmetric 6 × 6 matrix in Voigt notation.

$$\begin{bmatrix} \sigma_1 \\ \sigma_2 \\ \sigma_3 \\ \sigma_4 \\ \sigma_5 \\ \sigma_6 \end{bmatrix} = \begin{bmatrix} c_{11} & c_{12} & c_{13} & c_{14} & c_{15} & c_{16} \\ c_{21} & c_{22} & c_{23} & c_{24} & c_{25} & c_{26} \\ c_{31} & c_{32} & c_{33} & c_{34} & c_{35} & c_{36} \\ c_{41} & c_{42} & c_{43} & c_{44} & c_{45} & c_{46} \\ c_{51} & c_{52} & c_{53} & c_{54} & c_{55} & c_{56} \\ c_{61} & c_{62} & c_{63} & c_{64} & c_{65} & c_{66} \end{bmatrix} \begin{bmatrix} \varepsilon_1 \\ \varepsilon_2 \\ \varepsilon_3 \\ \varepsilon_4 \\ \varepsilon_5 \\ \varepsilon_6 \end{bmatrix} \quad (2)$$

Elastic constant matrix C_{ijkl} refers to the parameters that resist physical resilience when the material is subjected to the stress. Elastic constants reflect the macroscopic mechanical properties. Further reductions in the number of independent constants are possible by employing other symmetry considerations.

Elastic constants of single crystal can be determined by various methods, such as classical MD method calculation [49], the universal linear-independent coupling strains [50], first-principles methods [51, 52], and experimental measurements. For an aggregate crystalline structure, Voigt [53] proposed to find the average of stresses at all possible lattice orientations under a given strain, while Reuss [54] suggested to calculate the average of strains under a given stress instead. In addition to the experimental technique, there exist mainly two methods of calculating the elastic constants: (1) stress-strain curve relationship and (2) strain energy-strain relationship [55]. Elastic constants can be determined by the first-order derivative of the curve. Finally, the bulk and shear modulus of the crystalline phases can be estimated based on averaging methods.

The independent elastic coefficients corresponding to each crystal are listed in **Table 2**.

Crystal type	Lattice relations	C_{ij} number	and elastic coefficient	Method	Object
Cubic	a = b = c; α = β = γ = 90°	3	c_{11}, c_{12}, c_{44}	DFT	CaO, MgO
Hexagonal	a = b ≠ c; α = β = 90°, γ = 120°	5	c_{11}, c_{12}, c_{13}, c_{33}, c_{44}	DFT	Calcite, portlandite
				AFEM	Portlandite, graphene
Monoclinic	a ≠ b ≠ c; α = γ = 90°, β ≠ 90°	13	c_{11}, c_{22}, c_{33}, c_{12}, c_{13}, c_{23}, c_{44}, c_{55}, c_{66}, c_{15}, c_{25}, c_{35}, c_{46}	DFT	Tobermorite, gypsum
				MD	Tobermorite
Others	—	—	—	AFEM	CNT
				MD	C—S—H with C/S = 1.67

Table 3.
Independent elastic constants for typical crystals/structure considered.

As we can see in **Table 2**, the elastic properties of an isotropic material can be fully determined by calculating independent coefficients; thus, bulk modulus and shear modulus can be finally determined, using the relationship between the homogenized elastic moduli and elastic constants, and Young's modulus, shear modulus, the speed of sound, and other physical properties of the material.

Simulations of typical structures to determine their elastic moduli based on AFEM, DFT, and MD are shown in **Figure 13**.

Independent elastic constants for typical crystals/structure considered are in **Table 3**. A kind of AFEM method in Chapter 2 to simulate the previously proposed models at the atomic scale is shown in Chapter 2.

From **Table 3**, we can see that during the investigation of nanoscale modeling and mechanical properties, the DFT, AFEM, and MD method are especially concerned according to the modeling requirement of different crystal structures investigated. In summary, our research will provide further detailed parameters of our research objects and thus leads to a certain practical significance.

5. Mechanical properties calculation of typical structures for elastic properties

To get all 21 elements, only one of the six independent strains (ε_1, ε_2, ε_3, ε_4, ε_5, ε_6) nonzero each time. Using the average stress tensor, $< \sigma_{ij} >$, one can then define the effective elastic tensor by $< \sigma_{ij} > = < C_{ijkl} > \varepsilon_{ij}$, where ε_{ij} is the applied strain. Using this technique, all 21 elements can be evaluated. Methodology used for elastic constants evaluation is in **Table 4**.

In **Table 4**, the experimental technique is to build a microstructure using one of the above rules, assign elastic properties to the phases, and use the programs to compute the effective elastic moduli of the model.

Methodologies for elastic constant evaluation of considered structures are in **Table 5**.

Method	The method of measurement or calculation	Measured or calculated parameters
Experiment	Static, quasistatic method	Strain, stress
	Indentation method	Strain, stress
	Pulse-echo method	Velocity
	Resonance ultrasonic spectrum (RUS) method	Frequency
	X-ray or neutron method	Diffraction or scattering spectra
	Brillouin scattering method	Diffraction or scattering spectra
Calculation	Pending elasticity parameter method	Parameters from experiment
	Classical molecular dynamics	Energy, stress, strain
	Monte Carlo method	Energy, stress, strain
	First-principles molecular dynamics	Energy, stress, strain
	DFT-QHA method	Phonon density of states, Helmholtz free energy
(Present)	Atomic finite element method	Strain, stress

Table 4.
Methodology used for elastic constant evaluation.

Method	Measurement or calculation methods	Measured or calculated parameters
Experiment	Indentation method	Strain, stress
Classical numerical calculation	Classical molecular dynamics	Energy, stress, strain
	First-principles molecular dynamics	Energy, stress, strain
Method developed	AFEM	Stress, strain

Table 5.
Methodologies for the elastic constant evaluation of considered crystal structures.

From **Table 5**, it shows advantages and disadvantages of each modeling method. Besides, different typical structures are modeled using different modeling method to calculate mechanical properties. Researchers in materials science may be tasked with a number of challenging goals such as the development of new compounds or new molecular structure that may be faced with simply understanding and describing fundamental processes, explaining why one particular material is better than another. Modeling can address all of these challenges, provided that the method is fast, accurate, and works at the atomic scale.

5.1 AFEM method for C—C and hexagonal structures by ABAQUS

5.1.1 AFEM methodology for C—C and hexagonal structures

The AFEM/FEM linkage provides a seamless multiscale computation method for static analysis [56]. The AFEM element can be implemented into the ABAQUS finite element program with its USER-ELEMENT subroutine to solve several atomic-scale problems. To ensure that the AFEM/FEM multiscale computation method accurately represents the material behavior at both atomic and continuum scales, the continuum FEM elements should be based on the same interatomic potential as AFEM elements. The nonlinear spring element's definition of ionic bond or metallic bond can be calculated by MATLAB according to proper potential functions in various corresponding equilibrium conditions and then be imported into ABAQUS. For AFEM modeling, the calculation of U_r, U_θ, U_φ, and U_ω of Eq. (1) has been investigated in the reference [5], where the molecule's chemical energy is equivalent to the beam strain energy. The coupled field of van der Waals force between atoms or charged ions may be considered by means of the Lennard-Jones potential. Nonlinear spring elements can represent U_{vdw} potential.

5.1.2 The atomic finite element method modeling and ABAQUS

As a so-called atomic finite element method (AFEM) as a bridge between molecular dynamics and continuum mechanics has been proposed based on molecular mechanics method, where main force fields can be divided into two parts—covalent bond interaction and nonbonded force field. The nonlinear ABAQUS is a commercial finite element analysis program, which has been in use to analyze the structure in mechanical/material industry, civil engineering, and other field industries. The software is capable to analyze the stress and strain buildup in a variety of problems, to design the AFEM multiunit/element models. For defining nonlinear spring behavior, we can define nonlinear spring behavior by giving pairs of force-relative displacement values, which should be given in ascending order of relative displacement and should be provided over a sufficiently wide range of relative displacement values so that the behavior is defined correctly. Nonlinear spring is used to define force-relative displacement relations [57]. For crystal structure before homogenization, its elastic moduli in three directions are different, so is there a link of theoretical expression formula between axial elastic moduli and elastic coefficient? From Hooke's law, it can be inferred indirectly that axial elastic moduli are closely related to elastic coefficient. For mechanical properties, the anisotropy of the material shows the performance difference of axial moduli in three axial directions; formula between axial elastic moduli and elastic coefficient is derived in Chapter 2.

5.2 DFT method for typical structures by CASTEP and homogenization

Elastic constants can be calculated by CASTEP module using DFT method. For DFT method, the Hamiltonian and Kohn-Sham equation [58] as an approximation to simplification of Schrodinger equation can be used to solve the multiparticle systems. However, at a sufficiently fine scale, all materials are heterogeneous or can be said to be homogeneous, which introduces an important concept—homogenization.

5.2.1 The density functional theory calculation and CASTEP

The Hamiltonian and Kohn-Sham equations as an approximation are used for the simplification of Schrodinger equation to solve the multiparticle systems [59]. DFT includes not only the nucleus and electron kinetic energy term, but also interaction term of nucleus-nucleus and electron-electron [49]. CASTEP is an ab initio quantum mechanical program employing density functional theory (DFT) to simulate the properties of a wide range of solid materials at the atomic scale. CASTEP uses quantum mechanical calculations to study problems in chemicals and materials research, which is able to predict the structure of a material as well as many essential properties. In particular, it can predict electronic properties (e.g., band gaps and Schottky barriers), optical properties (e.g., phonon dispersion curves, polarizability, and dielectric constants), or physical properties such as elastic constants.

The computed elastic tensor of engineering material will have the symmetry of elastic constants matrix C_{ijkl} in general. One can analytically average this tensor, using the full C_{ijkl} form of the elastic moduli tensor, over the three Euler angles (α, β, γ) [60]. For solid material at macroscale, it can be seen that polycrystalline material is a collection of small grain, so elastic, plastic, and physical properties of the polycrystalline at macroscale are commonly thought to be relative with the mechanical properties of single crystals and polycrystalline themselves, including grain orientation and grain boundary structure.

First-principles calculations allow researchers to investigate the nature and origin of the electronic, optical, and structural properties of a system without the need for any experimental input. CASTEP is thus well suited to research problems in solid-state physics, materials science, chemistry, and chemical engineering where empirical models are lacking and experimental data may be sparse. In these areas, researchers can employ computer simulations to perform virtual experiments, leading to tremendous savings in costly experiments and shorter developmental cycles. Key procedure includes a transition state search algorithm that greatly facilitates determination of reaction profiles and energy barriers, essential to an understanding of kinetics. The full 6×6 tensor of the elastic constants can be predicted for a periodic structure of any symmetry. In the density functional theory part of the book, CASTEP is used to calculate the elastic constants of the single crystal structure, seen in Chapter 3.

5.2.2 Homogenized moduli by Y-parameter method based on elastic constants

Although single crystals generally exhibit anisotropic mechanical behaviors, macroscopic averaging assuming a randomly distribution of initial crystals leads to isotropic mechanical properties such as Young's modulus. Polycrystalline material is composed of single crystal, the probability distribution depicting upward almost the same as in three-dimensional space, resulting in an isotropic behavior of mechanical properties at macro-scopic scale. Meanwhile, Voigt model based on the assumption that all grains have the same deformation strain is given, by which the upper bound of polycrystalline and elastic constitutive relation can be given, while the Reuss model is based on the hypothesis of the same stress for polycrystalline, where the lower bound of polycrystalline and another form of elastic constitutive relation can be given. However, the constitutive relation by Voigt and Reuss model is not always accurate, especially for the lower bound. Y-parameter of cubic structure was firstly introduced and presented by Zheng and Min [61, 62] in 2009. Based on the elastic constants of cubic crystal, the corresponding Y-parameter, as new evaluation criteria, has been proposed. Elasticity of single crystal and mechanical properties of polycrystalline material has been closely integrated. By comparing various calculations method to determine homogenized moduli of the polycrystalline material composed of a single crystal, for example, the certain stress of Reuss model [54] and the certain strain of Voigt model [53], the Y-parameter, in the theoretical calculation to forecast the elastic modulus of polycrystalline material, has a high consistency. So our work aims to complete this part of the theoretical analysis for polycrystalline, with comparison of elastic constants measured by references.

This interesting Y-parameter for cubic crystals can be used to determine the homogenized moduli of cubic structures. Moreover, the structural and the elastic properties of CH and calcite are investigated by ab initio plane-wave pseudopotential density functional theory method in Chapter 3, and then the credibility of Y parameters for determining elastic moduli of CH and calcite structures is proved. We will then describe these interesting Y parameters of cubic and hexagonal polycrystalline in detail in Chapter 4. In fact, due to the difficulty to accurately measure elastic constants of each crystal for polycrystalline, the fourth-order tensor connections between mean stress and mean strain called the elastic constitutive relation are impossible to be determined by numerical integration method. Then elastic moduli (Young's, shear, and bulk moduli) can be obtained by means of Voigt and Reuss bounds, which are based on the C_{ij} calculated. As the complication of the monoclinic crystals (11 Å tobermorite and gypsum), elastic moduli of monoclinic crystals will be still calculated by the classical Voigt-Reuss-Hill estimation, in Chapter 4.

5.3 MD method and indentation simulation for C—S—H by LAMMPS

5.3.1 Elastic moduli by stress-strain curve slope method using MD

To simulate large-scale structures instead of these small crystals by DFT, the corresponding potentials were implemented in large-scale atomic/molecular massively parallel simulator (LAMMPS) code [63]. Note that LAMMPS is an open-source code and thus can be actively modified by users as well as the developers. At the end of each time interval, a new box length is computed. The elastic modulus of typical structures can be approximated and computed from the slope of the linear portion of the stress-strain curve.

In the molecular dynamic part of this book, LAMMPS (large-scale atomic molecular massively parallel simulator) is a classical MD code used in this study. LAMMPS is developed by Steve Plimpton and his group of Sandia National Laboratories, which can be used to calculate the tension, compression, and shear deformation processes of the large-scale supercell structures. An input file was generated for LAMMPS code to calculate the response of C—S—H structures to uniaxial loading, which leads to strain-stress data to calculate elastic moduli we aim to obtain in Chapter 5.

5.3.2 Elastic moduli by nanoindentation simulation

Mechanical properties of concrete as a composite material depend on microstructure and even depend on properties of crystal of each paste at nanoscales. Concrete is a complex heterogeneous material whose mechanical properties can vary substantially from point to point. Nanomechanics and nanotechnology are used to explore the composition, behavior, and nanomechanical properties of cement paste phases with submicron resolution. Nanoindentation technique bridges the gap between atomic force microscopy (AFM) and macroscale mechanical testing. The corresponding load-depth curve can be measured directly by nanoindentation simulation; thus, elastic modulus can be determined.

Therefore, nanoindentation technique as a useful tool can be used to better understand the submicron mechanical properties of cement pastes. Because of its small probe size, indentation tools are previously mainly used to measure local material properties in small, thin, and heterogeneous materials. Nanoindentation simulation becomes a hot spot in recent years; based on continuum mechanics, nanoindentation simulation can reveal atomic deformation mechanism under a indent force; the smallest unit with embedded atomic force is used in the indentation area of the multiscale model. Some methods or programs can be used partially to design and analyze multiscale model, which includes the finite element method, the boundary element method, finite difference method, and the discrete element method. ABAQUS can be used to program and simulate the nanoindentation process even with the nanoscale unit. The nanoindentation model with the unit of a nanometer size will be simulated and discussed in Chapter 6.

6. Conclusions

The development of modeling tools to describe and predict the mechanical properties of structural materials proves an undeniable practical importance. It is now well recognized that such an objective can be achieved through the linking of the structure of materials at the microscopic nanoscale or with their performance. Nanoscale modeling and mechanical properties by using the density functional theory, a so-called atomic finite element method, and the classical molecular dynamics methods are especially concerned in this work. Such tools will be more accurate and they can relate the structure of materials at the atomic or microscopic scale with their performance. In quantum mechanics field, when applied to real systems including several atoms and electrons, a proper solution of the stationary equation remains a challenge due to its complexity. However, the mean field theory (electrons move as independent particles in an effective potential generated by the ions and the other electrons) is also effective. Especially for molecular dynamics simulations, the model scale can be greatly extended.

In addition to the tradition of DFT methods and MD, a methodology for AFEM modeling at the atomic scale (or nanoscale) is explored since the macroscopic properties depend largely on the physicochemical properties of the interatomic bonds. This book is committed to model and to calculate mechanical properties (especially Young's modulus) of the different typical anisotropic crystal structures by programming using three methods above. It therefore connects nanoscale modeling and continuous pattern of deformation behavior by identifying relevant parameters from small scales to larger scales. Typical crystal structures will be discussed systematically in this book to determine their elastic moduli.

In summary, such modeling tools will be more accurate and they can relate the structure of materials at the atomic or microscopic scale with their performance. It is not far for us to deeply investigate mechanical properties with the development of computer hardware level, the interatomic potential functions accurately described, new high-efficiency algorithms, etc.

Acknowledgements

The author acknowledges the financial support provided by the China Scholarship Council and the startup foundation of Xi'an Shiyou University. Thanks Qiufeng Wang for her proofreading, and very sincere thankness to prof. Fabrice Bernard and prof. Siham Kamali-Bernard for the their enthusiastic and responsible guideness.

Author details

Jia Fu[1,2]

1 Xi'an Shiyou University, Xi'an, China

2 INSA de Rennes, Rennes, France

*Address all correspondence to: fujia@xsyu.edu.cn

References

[1] Raabe D. Computational Materials Science: The Simulation of Materials Microstructures and Properties. Weinheim: Wiley-VCH; 1998

[2] Caddell H, Hosford WF, Caddell RM. Metal Forming. 2nd ed. Englewood: Prentice Hall; 1993

[3] Nye JF. Physical Properties of Crystals: Their Representation by Tensors and Matrices. Oxford: Oxford University Press; 1985

[4] Huang Y. A User-Material Subroutine Incroporating Single Crystal Plasticity in the ABAQUS Finite Element Program. Cambridge(Massachusetts): Harvard University; 1991

[5] Li C, Chou T-W. A structural mechanics approach for the analysis of carbon nanotubes. International Journal of Solids and Structures. 2003;**40**: 2487-2499

[6] Faucon P, Delaye JM, Yirlet J, et al. Study of the structural properties of the C-S-H (I) by molecular dynamics simulation. Cement and Concrete Research. 1997;**27**(10):1581-1590

[7] Jia F, Huai S. An ab initio force field for predicting hydrogen storage in IRMOF materials. Journal of Physical Chemistry C. 2009;**113**(52):21815-21824

[8] Arash B, Wang Q. A review on the application of nonlocal elastic models in modeling of carbon nanotubes and Graphenes. Computational Materials Science. 2012;**51**:303-313

[9] Perdew JP, Burke K. Comparison shopping for a gradient-corrected density functional. International Journal of Quantum Chemistry. 1996;**57**(3): 309-319

[10] Raabe D. Challenges in computational materials science. Advanced Materials. 2002;**14**(9): 639-650

[11] Schrodinger E. An undulatory theory of the mechanics of atoms and molecules. Physical Review. 1926;**28**(6): 1049-1070

[12] Levine IN. Quantum Chemistry. 4th ed. Englewood Cliffs, N.J: Prentice Hall; 1991. 629 p

[13] Handy NC, Pople JA, Shavitt I. Samuel Francis boys. Journal of Physical Chemistry. 1996;**100**(15):6007-6016

[14] Car R, Parrinello M. Unified approach for molecular dynamics and density-functional theory. Physical Review Letters. 1985;**55**(22):2471-2474

[15] Kohn W, Sham LJ. Self-consistent equations including exchange and correlation effects. Physical Review. 1965;**140**(4A):A1133

[16] Marx D, Hutter J. Ab Initio Molecular Dynamics Basic Theory and Advanced Methods. Cambridge: Cambridge University Press; 2009

[17] Alder BJ, Wainwright TE. Molecular dynamics by electronic computers. In: International Symposium on Statistical Mechanical Theory of Transport Processes. Brussels: Interscience, Wiley; 1956

[18] Lisi AG. An Exceptionally Simple Theory of Everything. 2007. arXiv: 0711.0770[hep-th]

[19] Neugebauer R, Wertheim R, Semmler U. The atomic finite element method as a bridge between molecular dynamics and continuum mechanics. Journal of Multiscale Modelling. 2011;**3** (01–02):39-47

[20] Rappe AK, Casewit CJ, et al. UFF: A full periodic-table force-field for

molecular mechanics and molecular dynamics simulations. Journal of American Chemical Society. 1992;**114**: 10024-10035

[21] Benboudjema F, Torrenti JM. Early-age behaviour of concrete nuclear containments. Nuclear Engineering and Design. 2008;**238**(10):2495-2506

[22] Randall N. Mechanical properties of cementitious materials. CSM Instruments Application Bulletin. 2009; **29**:1-4

[23] Manzano MH. Atomistic Simulation Studies of the Cement Paste Components. Bilbao: Servicio Editorial de la Universidad del País Vasco/Euskal Herriko Unibertsitatearen Argitalpen Zerbitzua; 2014

[24] Zadeh VZ, Bobko CP. Nanomechanics and Multiscale Modeling of Sustainable Concretes. Raleigh: North Carolina State University; 2013

[25] Locher FW. Die festigkeit des zements. Beton. 1976;**26**(8):283-285

[26] Regourd M. L'action de l'eau de mer sur les ciments. Annales de l'Institut Technique du Bâtiment et des Travaux Publics. 1975;**25**:86-102. (In French)

[27] Goto S et al. Composition and morphology of hydrated tricalcium silicate. Journal of the American Ceramic Society. 1976;**59**(7–8): 281-284

[28] Richardson IG. Tobermorite/jennite- and tobermorite/calcium hydroxidebased models for the structure of C-S-H: Applicability to hardened pastes of tricalcium silicate, beta-dicalcium silicate, Portland cement, and blends of Portland cement with blast-fumace slag, metakaolin, or silica fume. Cement and Concrete Research. 2004;**34**(9):1733-1777

[29] Richardson IG, Groves GW. Microstructure and microanalysis of hardened ordinary Portland cement pastes. Journal of Materials Science. 1993;**28**(1):265-277

[30] Loukili A. Self Compacting Concrete. New Jersey: John Wiley & Sons; 2013

[31] Berger M, et al. Cement nanotubes as a natural means for reinforcing concrete. 2012; Available from: http://www.nanowerk.com/Michael_Berger.php

[32] Ma S, Scheider I, Bargmann S. Anisotropic constitutive model incorporating multiple damage mechanisms for multiscale simulation of dental enamel. Journal of the Mechanical Behavior of Biomedical Materials. 2016;**62**:515-533

[33] Zahn D. Multi-scale simulations of apatite–collagen composites: From molecules to materials. Frontiers of Materials Science. 2017;**11**(1):1-12

[34] Sagadevan S, Pal K, Chowdhury ZZ. Scalable synthesis of CdS–graphene nanocomposite spectroscopic characterizations. Journal of Materials Science: Materials in Electronics. 2017; **28**(22):17193-17201

[35] Thirugnanasambandan T, Pal K, Sidhu A, et al. Aggrandize efficiency of ultra-thin silicon solar cell via topical clustering of silver nanoparticles. Nano-Structures & Nano-Objects. 2018;**16**: 224-233

[36] Sagadevan S, Pal K, Chowdhury ZZ, et al. Controllable synthesis of graphene/ZnO-nanocomposite for novel switching. Journal of Alloys and Compounds. 2017;**728**:645-654

[37] Bargmann S, Scheider I, Xiao T, et al. Towards bio-inspired engineering materials: Modeling and simulation of the mechanical behavior of hierarchical bovine dental structure. Computational Materials Science. 2013;**79**:390-401

[38] Pal K, Mohan MLNM, Foley M, et al. Emerging assembly of ZnO-nanowires/graphene dispersed liquid crystal for switchable device modulation. Organic Electronics. 2018; **56**:291-304

[39] Humburg H, Volkmann E, Koch D, et al. Combination of biological mechanisms for a concept study of a fracture-tolerant bio-inspired ceramic composite material. Journal of Materials Science. 2014;**49**(23):8040-8050

[40] Fratzl P, Weinkamer R. Nature's hierarchical materials. Progress in Materials Science. 2007;**52**(8):1263-1334

[41] Gao H, Ji B, Jäger IL, et al. Materials become insensitive to flaws at nanoscale: Lessons from nature. Proceedings of the National Academy of Sciences. 2003;**100**(10):5597-5600

[42] Kamali-Bernard S, Bernard F. Effect of tensile cracking on diffusivity of mortar: 3D numerical modelling. Computational Materials Science. 2009; **47**:178-185

[43] Bernard F, Kamali-Bernard S, Prince W. 3D multi-scale modelling of mechanical behavior of sound and leached mortar. Cement and Concrete Research. 2008;**38**:449-458

[44] Koenders E, Schlangen E, Breugel K. Multi-scale Modelling: The Delft code. Conference of CONMOD08 in Delft University of Technology. 2008;**1**: 1-8

[45] Eftekhari M, Mohammadi S. Molecular dynamics simulation of the nonlinear behavior of the CNT-reinforced calcium silicate hydrate (C–S–H) composite. Composites Part A: Applied Science and Manufacturing. 2016;**82**:78-87

[46] Tastani SP, Konsta-Gdoutos MS, Pantazopoulou SJ, Balopoulos V. The effect of carbon nanotubes and polypropylene fibers on bond of reinforcing bars in strain resilient cementitious composites. Frontiers of Structural and Civil Engineering. 2016; **10**(2):214-223

[47] Raki L, Beaudoin JJ, Alizadeh R, Makar JM, Sato T. Cement and concrete nanoscience and nanotechnology. Materials. 2010;**3**:918-942

[48] Makar JM, Chan GW. Growth of cement hydration products on single-walled carbon nanotubes. Journal of the American Ceramic Society. 2009;**92**(6): 1303-1310

[49] Benazzouz BK. Etude par dynamique moléculaire des propriétés mécaniques et thermodynamiques de l'argile de type kaolinite [Thèse de doctorat]. France: l'université Lille Sciences et Technologies; 2010. (In French)

[50] Yu R, Zhu J, Ye HQ. Calculations of single-crystal elastic constants made simple. Computer Physics Communications. 2010;**181**:671-675

[51] Page YL, Saxe P. Symmetry-general least-squares extraction of elastic data for strained materials from ab initio calculations of stress. Physical Review B. 2002;**65**:104104

[52] Perger WF, Criswell J, Civalleri B, et al. Ab-initio calculation of elastic constants of crystalline systems with the crystal code. Computer Physics Communications. 2009;**180**:1753-1759

[53] Voigt W. Lehrbuch der Kristallphysik Teubner. Leipzig 1910; reprinted (1928) with an additional appendix. Leipzig, Teubner, New York, Johnson Reprint

[54] Reuss A. Berechnung der fließgrenze von mischkristallen auf grund der plastizitätsbedingung für einkristalle. ZAMM—Journal of Applied Mathematics and Mechanics/Zeitschrift

für Angewandte Mathematik und Mechanik. 1929;**9**(1):49-58

[55] Bin W, Tianjiao S. Method for Computing High-Temperature Elastic Coefficient of Any Crystal System Material. C.N. Patent [P/OL] CN102353692(B); 2013

[56] Liu B, Huang Y, Jiang H, Qu S, Hwang KC. The atomic-scale finite element method. Computer Methods in Applied Mechanics and Engineering. 2004;**193**(17):1849-1864

[57] Abaqus Users Manual. Version 6.13–2, Internet Manual of Simulia. Providence, Rhode Island, USA: Dassault Systémes Simulia Corp.; 2013

[58] Bauernschmitt R, Ahlrichs R. Stability analysis for solutions of the closed shell Kohn–Sham equation. The Journal of Chemical Physics. 1996; **104**(22):9047-9052

[59] Aaeid A. Études ab initio et dynamique moléculaire des propriétés structurales et thermodynamiques de la calcite et la witherite sous hautes pressions [PhD thesis]. France: l'Université des Sciences et Technologies de Lille; 2010. (In French)

[60] Arfken GB. Mathematical Methods for Physicists. Chapter on Rotation Matrices. New York: Academic Press; 2013

[61] Zheng LIN, Min L. The elastic constants of polycrystalline materials with cubic system structural single crystals. Acta Physica Sinica. 2009; **58**(6):4096-4102

[62] Zheng L, Min L. Y elastic constants of polycrystalline materials with hexagonal system structure. Acta Physica Sinica. 2009;**58**(12):8511-8521

[63] Plimpton S, Thompson A, Crozier P. Molecular Dynamics Simulations from SNL's Large-scale Atomic/Molecular Massively Parallel Simulator (LAMMPS). 2007; Available from: http://lammps.sandia.gov

Chapter 2

The Advanced Atomic Finite Element Method: Modeling and Application

Jia Fu

Abstract

An atomic finite element analysis is developed in this chapter. At atomic scale, the interatomic bonding forces of van der Waals and the covalent chemical bond are taken into account. In addition to the tradition atomistic simulation methods, the more specific objective of this chapter is to develop a modeling methodology at the atomic scale (or nanoscale) since the macroscopic properties largely depend on the physicochemical properties of the interatomic bonds. Above all, the methodology is applied to study the behavior of carbon nanotubes, whose development has experienced strong growth in recent years and can be used for quality mechanical reinforcement. These carbon nanotubes are formed by repeating zigzag carbon-carbon bonds. Development of atomic finite element method (AFEM) can be traced back to the homogenized elastic properties of various graphene structures (single-layer graphene sheet, zigzag single-walled carbon nanotubes, triple-layer graphene sheet). Moreover, the AFEM is developed to investigate the Young modulus of one of the main constituents of cement-based materials, portlandite (CH). Besides, a theoretical formula between axial modulus (Ex, Ey, and Ez) and elastic constants of hexagonal crystal is derived by crystal elastic theory. Modeling of single crystal can be traced back to the homogenized elastic properties of polycrystals.

Keywords: Portlandite, AFEM, DFT, anisotropy, Young's modulus

1. Introduction

Composite material reinforced by carbon nanotubes (CNT) has a high strength, mechanical shock, thermal shock resistance, and fracture toughness, thus has a potential application with greatly improved properties. The "ene aluminum alloy"-graphene-reinforced aluminum matrix nanocomposites are firstly discovered for high properties in recent years. Carbon nanotube (CNT) and graphene have established a natural link; graphene was firstly discovered and separated by Geim [1] and the influential study shows that CNT can be formed by rolling a cylindrical graphene [2]. CNT has caused a widespread concern in civil engineering application. Due to disadvantages of lower tensile strength and lower impact strength of cement-based material, CNT as a reinforcing matrix material has higher mechanical stability properties and impact resistance. The difficulty in seeking the analytical solution of CNT Young's modulus is existing due to choosing various wall thicknesses, zigzag or armchairs, *D*/*L* ratio, indirect measurements, dimensions, and modeling methods, etc. Young's modulus of single-walled carbon nanotubes (SWCNT) is about 1.5–5.0 TPa [3] for individual CNT, 1.033–1.042 TPa [4] with thickness 0.34 nm by using nanoscale continuum mechanics method (CM), 5.5 TPa for SWCNT based on MD method with thickness 0.066 nm by Yakobson et al. [5], 4.88 TPa for SWCNT based on CM by Vodenitcharova and Zhang [6], 5.1 TPa for CNT based on the electronic energy band theory (EEBT) by Zhou et al. [7] with considering the total energy of all electrons. There is no agreement of Young's modulus of CNT and graphene in the analytical method up to nowadays; the disagreement of Young's modulus of CNT and graphene is actually due to the restriction of both the theory and experiment. Especially for graphene, atomic van der Waals forces between the layers cannot be ignored. As the measurement of a single CNT and graphene has some difficulty especially for a certain type (zigzag type or air-chairs type), it therefore causes our interest to present the exact forms of modeling at nanoscale.

It is common that graphene and nanotube structures sharing the same C—C bond have a wide potential application. CNT has similar properties of fullerene molecules; its high strength has caused a widespread concern in various application fields. Graphene is considered to be an atomic crystal with flat polycyclic aromatic hydrocarbons (the structure is very stable). It can be rolled into a barrel-shaped carbon nanotube [2], showing the similar properties of nanotube and fullerene molecule. However, the theoretical mechanical deformation analysis is so complicated that the nonlinear analytical method is needed where van der Waals force should be considered and the corresponding nonlinear transformation is needed.

For the value of Young's modulus of carbon nanotubes (CNT), after the separation of graphene by Geim [1], Young's modulus is 1.033–1.042 TPa [8] with thickness 0.34 nm by using nanoscale continuum mechanics method, then 1.5–5.0 TPa for CNT by Treacy et al. [3], 5.5 TPa for SWCNT based on MD method with thickness 0.066 nm by Yakobson et al. [5], 4.88 TPa for SWCNT based on continuous mechanics method by Vodenitcharova and Zhang [6], 5.1 TPa for CNT based on the electronic energy band theory by Zhou et al. [7] with considering the total energy of all electrons. Researches of CNT are mainly listed in **Table 1**.

From **Table 1**, the Young's moduli of CNT with considering various thicknesses using various methods are not all the same. Actually, there is no agreement of Young's modulus nowadays due to choosing various wall thickness, zigzag or armchair, D/L ratio, indirect measurements, dimensions, methods, etc.

Thus, a new so-called atomic finite element method (AFEM) as a bridge between molecular dynamics and continuum mechanics [22] can be used to investigate elastic properties of covalent bond structures. In 2011, the AFEM [22] is proposed, in which the interatomic bonds are modeled as nonlinear spring elements. Besides, for the typical C—C frame-like structures (e.g., graphene and SWCNT), a structural mechanics approach proposed by Li and Chou [8] has good accuracy and stability to calculate their Young's modulus, and the modeling part of covalent bond can be modeled as an equivalent beam element. As a structural mechanics approach, Li and Chou [8] have successfully modeled the C—C covalent bond structures by equivalent beam elements and a so-called AFEM [22] has solved the interatomic bonds modeled by nonlinear spring elements, modeling of nanostructure realized by an equivalent beam and nonlinear spring elements becomes possible.

In this chapter, carbon nanotubes (CNT), hexagonal graphene, and portlandite (CH) are particularly investigated. CNT, has its unique physical and chemical characteristics, has higher mechanical stability properties and impact resistance, thus can be used as a reinforcing material or composite material matrix. CH is a typical constituent of cement paste and its Young's moduli are needed in the

Year	Author	Method	*t* (nm)	*E* (TPa)	*E.t*
1996	Yakobson et al. [5]	MD	0.066	5.5	0.3630
1999	Goze et al. [9]	Tight binding	0.34	1.26	0.4284
2000	Popov et al. [10]	FCM	0.33	1.02	0.3368
2000	Zhou et.al [7]	Tight-binding MD	0.074	5.1	0.3774
2001	Kudin et al. [11]	Ab initio	0.089	3.859	0.3450
2002	Tu and Ou-Yang [12]	LDA	0.075	4.7	0.3525
2003	Li and Chou [8]	Structure mechanics	0.34	1.02	0.3463
2003	Chang and Gao [13]	Molecular mechanics	0.36	0.997	0.3592
2004	Sears and Batra [14]	Molecular mechanics	0.134	2.52	0.3377
2004	Pantano et al. [15]	Continuum shell	0.075	4.84	0.3630
2005	Zhang et al. [16]	C-B rule	0.335	1.08	0.3618
2006	To [17]	FEM	0.34	1.024	0.3482
2007	Chandraseker and Mukherjee [18]	Ab initio	0.335	0.99	0.3317
2008	Wang and Zhang [19]	Continuum shell	0.1	3.5	0.350
2008	Jinan Lu and Chen [20]	FEM	0.34	1.274	0.4332
2010	Mahmood et al. [4]	Continuum mechanics	0.34	1.033–1.042	0.3512–0.3543
2012	Xiaoxing and Zhong Hu [21]	Equivalent-continuum	0.33	1.058	0.349

Table 1.
Elastic modulus and wall thickness of nanotubes.

MuMoCC platform. Above all, Young's modulus formula of zigzag CNT can be obtained and then the AFEM method is verified. Modeling of TLGSs structure and CH crystal using the equivalent beam-spring element is also investigated. This chapter aims to develop a modeling method of nanostructure realized by an equivalent beam and nonlinear spring elements, of which the nonlinear spring element's definition of nonbonding can be realized according to a kind of Lennard-Jones potential in various corresponding equilibrium conditions.

2. Modeling methods of continuum mechanics and AFEM modeling

2.1 Classical modeling methods by continuum mechanics method

2.1.1 Classical molecular mechanics method (MM)

Generally, the force field is expressed in the form of steric potential energy. It depends solely on the positions of the nuclei constituting the molecule. The total steric potential energy, omitting the electrostatic interaction, is a sum of energies due to valence or bonded interactions and nonbonded interactions [23]:

$$U = \sum U_r + \sum U_\theta + \sum U_\phi + \sum U_\omega + \sum U_{vdw} \tag{1}$$

where U_r is for a bond stretch interaction, U_θ for a bond angle bending, U_ϕ for a dihedral angle torsion, U_ω for an improper (out-of-plane) torsion, U_{vdw} for a nonbonded van der Waals interaction. Interatomic interactions in molecular mechanics by Li and Chou are shown in **Figure 1**.

Li and Chou have proposed the molecular mechanics method (MM) [8], under the small deformation premise, to simplify the calculation, the simplest harmonic forms to merge the dihedral angle torsion and the improper torsion into a single equivalent term are adopted, which are as follows:

$$U_r = \frac{1}{2}k_r(r - r_0)^2 = \frac{1}{2}k_r(\Delta r)^2 \tag{2}$$

$$U_\theta = \frac{1}{2}k_\theta(\theta - \theta_0)^2 = \frac{1}{2}k_\theta(\Delta\theta)^2 \tag{3}$$

$$U_\tau = U_\phi + U_\omega = \frac{1}{2}k_\tau(\Delta\phi)^2 \tag{4}$$

where k_r, k_θ, and k_τ are the bond stretching force constant, bond angle bending force constant, and torsional resistance, respectively, and the symbols Δr, $\Delta\theta$, and $\Delta\phi$ represent the bond stretching increment, the bond angle change, and the angle change of bond twisting, respectively.

Assuming that the sections of carbon-carbon bonds are uniformly circular, thus $I_x = I_y = I$. So parameters of EA, EI, and GJ can be determined. As the molecules' chemical energy is equivalent to the strain energy of beam, thus, there are the following approximate relationships:

$$\frac{EA}{L} = k_r, \quad \frac{EI}{L} = k_\theta, \quad \frac{GJ}{L} = k_\tau \tag{5}$$

where L is the beam length of the C—C bond, EA represents tensile stiffness, EI represents bending stiffness, GJ represents torsional stiffness of equivalent beam, k_r is the stretching constant, k_θ is the bending constant, and k_τ is the torsional constant. Obviously, these formulas have established the equivalent relation between the macrobeam structure and microscopic molecular structure. As long as the force constants k_r, k_θ, and k_τ are known, the sectional stiffness parameters EA, EI, and GJ can be readily obtained. Then, a mathematical modeling—AFEM modeling—of ionic crystals is proposed by means of interatomic potentials.

2.1.2 The chemical bond element method

The chemical bond element method has been proposed to link the relation between finite element method and continuum mechanics considering atomic

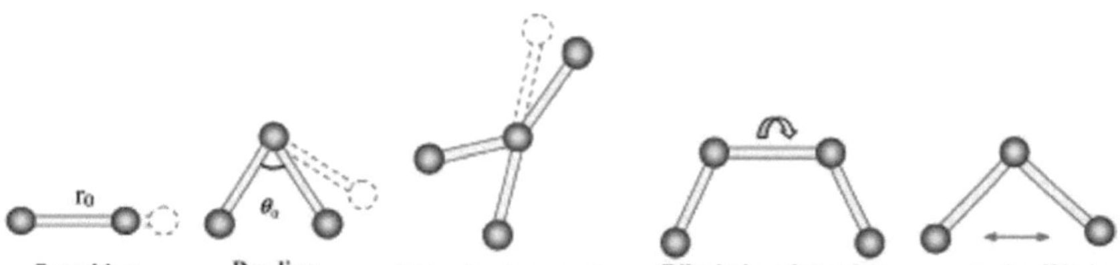

Figure 1.
Interatomic interactions in MM by Li and Chou.

interaction, where the local geometry of chemical bonds of most materials is characterized by a bond length r, bend angle θ, and dihedral angle φ, shown in **Figure 2(a)**.

Chemical bond element nodal forces and displacements are shown in **Figure 2(b)**. As is shown in **Figure 2(b)**, there are three nodal forces for each node: F_{xi}, F_{yi}, M_i for node i and F_{xj}, F_{yj}, M_j for node j. There are three nodal displacements for each node: u_i, v_i, θ_i for node i and u_j, v_j, θ_j for node j.

The nodal force increment and the displacement increment in each loading step in the x-direction are:

$$\Delta F_{xi} = k_r(\Delta u_i - \Delta u_j) \tag{6}$$

$$\Delta F_{xj} = k_r(-\Delta u_i + \Delta u_j) \tag{7}$$

$$\Delta M_i = -2k_{\theta i}(\Delta\theta_L - \Delta\theta_i) \tag{8}$$

$$\Delta M_j = 2k_{\theta j}(\Delta\theta_j - \Delta\theta_L) \tag{9}$$

where $\Delta\theta_L = \frac{\Delta v_j - \Delta v_i}{L}$, k_r is the stretching stiffness, k_{yi} and k_{yj} are the bending stiffness of joints i and j, and L is the length of the element.

For the C—C bonds, parameters by references [8, 24] are as follows: $k_r = 2.78\,\text{aJ}/\mathring{\text{A}}^2$, $k_\theta = 0.498\,\text{aJ/rad}$, $\varepsilon = 0.585 \times 10^{-3}\text{aJ}$, $r_0 = 1.47 \times 10^{-10}\text{m}$, $\theta_0 = 1.88\,\text{rad}$, $\sigma = 3.53 \times 10^{-10}\text{m}$, $k_r = 938\,\text{kcal} \times \text{mol}^{-1} \times \mathring{\text{A}}^{-2}$, $k_\theta = 126\,\text{kcal} \cdot \text{mol}^{-1} \cdot \text{rad}^{-2}$, $k_\tau = 40\,\text{kcal} \cdot \text{mol}^{-1} \cdot \text{rad}^{-2}$, $a_{C-C} = L = 0.142\,\text{nm}$. For the round section of C—C bonds, when t = 0.34 nm, thus A = 0.090792 (nm^2), E = 1019.259 (GPa), and I = 1.21959 × 10^{-4} (nm^4).

2.2 AFEM realization at nanoscale considering atomic interaction

Continuum methods (such as FEM) use averaged description for modeling the material properties. As a result, they are not capable of modeling phenomena at the atomic scale. Localized nonlinear deformation, defects, nanoscale materials, and structures (e.g., CNTs) as well as various nanoscale phenomena need to be modeled with the atomic scale accuracy. CNT types and characteristic parameters are classified in **Table 2**.

Where chiral vector $\vec{C_h} = n\vec{a_1} + m\vec{a_2}$, $a_1 = a_2 = \sqrt{3}a_{c-c}$, so $C_h = \sqrt{3}a_{c-c}\sqrt{n^2 + nm + m^2}$; chiral angle $\cos\theta = \frac{2n+m}{2\sqrt{n^2+nm+m^2}}$ or $\theta = tg^{-1}\left[\frac{\sqrt{3}m}{m+2n}\right]$; diameter $d = \frac{C_h}{\pi} = \frac{\sqrt{3}}{\pi}a_{cc}\sqrt{n^2 + nm + m^2}$.

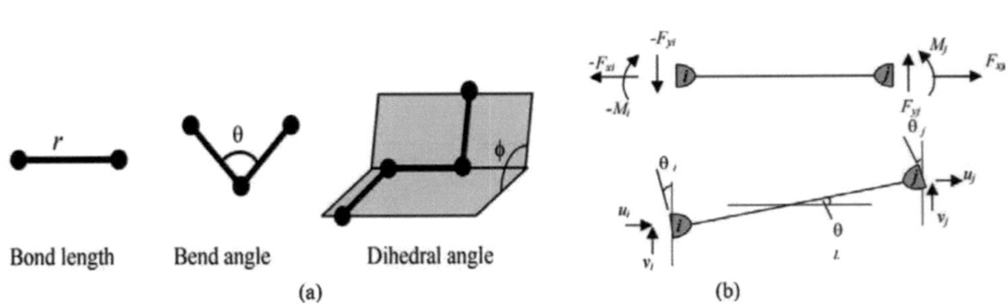

Figure 2.
Geometry of chemical bonds, nodal forces, and nodal displacements. (a) Geometry of chemical bonds. (b) Nodal forces and nodal displacements.

Types	Cross-sectional shape of carbon ring	Chiral angle θ	Chiral vector C_h
Armchair		30°	(n,n)
Zigzag		0°	(n,0)
Chiral type	Due to the chiral angle varies	$0° < \lvert\theta\rvert < 30°$	(n,m)

Table 2.
Different types, chiral angle, and chiral vector of CNT.

2.2.1 AFEM modeling based on molecular mechanics

Molecular mechanics method mainly includes two basic assumptions. The first assumption is the Born-Oppenheimer approximation, of which the electronic function is centered on the nuclear positions, with the electrons around them in an optimal distribution [25]. The second assumption is that each nucleus is seen as a classical particle. Electrons are not taken into account and atoms are simulated as spheres, with an assigned radius and charge, interacting by means of a collection of interatomic potentials [26, 27].

The AFEM is proposed based on the development of nanotechnology across multiple length scales [22]. The AFEM can be used to calculate the van der Waals force in the plane. For a system composed of N atoms, the total energy of the system can be expressed as a sum of all the chemical energy stored:

$$E_{total}(x) = U_{total}(x) - \sum_{i=1}^{N} \overline{F}_i \cdot x_i \tag{10}$$

where x_i is the coordinate of atom i vector and $\overline{F}_i$ is the external force exerted on the atom i vector. The minimum energy state is:

$$\frac{\partial E_{total}(x)}{\partial x} = 0 \tag{11}$$

For $\partial F_{total}(x)$, the initial state $x^{(0)}$ can be expanded by the developed Taylor, we obtain:

$$E_{total}(x) \approx E_{total}\left(x^{(0)}\right) + \left.\frac{\partial E_{total}(x)}{\partial x}\right|_{x=x^{(0)}} \cdot \left(x - x^{(0)}\right) + \frac{1}{2}\left(x - x^{(0)}\right)^T \left.\frac{\partial^2 E_{total}(x)}{\partial x \partial x}\right|_{x=x^{(0)}} \left(x - x^{(0)}\right) \tag{12}$$

Each atom not only forms a direct bond with its three adjacent atoms, but also has an interaction with its six subadjacent atoms. Newton-Raphson iteration is as follows:

$$\psi(x^n) = -\left(\frac{dE_{total}}{dx}\right)_n + \overline{F} \tag{13}$$

$$\psi\left(x^{n+1}\right) = \psi(x^n) + \left(\frac{d\psi}{dx}\right)_n \Delta x^n = 0 \tag{14}$$

According to Eq. (14), we can obtain:

$$\Delta x^n = -\left(\frac{d\psi}{dx}\right)^{-1} \cdot \psi(x^n) \tag{15}$$

$$x^{n+1} = x^n + \Delta x^n \tag{16}$$

where $\overline{F}$ is the external force vector, $\psi(x^n)$ is imbalance force vector after n iterations, x^n is the vector of generalized coordinates of atoms after n iterations. Each iteration needs to compute the total potential energy of the system and coordinates of the first and the second derivatives.

For the expression of K and F, the corresponding expressions are as:

$$K = -\frac{d\psi}{dx} = \frac{d^2 E_{total}}{dx^2} \tag{17}$$

$$F = -\frac{dE_{total}}{dx} + \overline{F}. \tag{18}$$

The above equation can be written as:

$$K \cdot u = P \tag{19}$$

For the atoms before deformation, the expression can be described as follows:

$$K = \left.\frac{\partial^2 E_{total}}{\partial x \partial x}\right|_{x=x^{(0)}} = \left.\frac{\partial^2 U_{total}}{\partial x \partial x}\right|_{x=x^{(0)}} \tag{20}$$

$$P = \left.\frac{\partial E_{total}}{\partial x}\right|_{x=x^{(0)}} = \overline{F} - \left.\frac{\partial U_{total}}{\partial x}\right|_{x=x^{(0)}} \tag{21}$$

K and P can be directly obtained by the interatomic interaction potential. The appropriate boundary conditions and finite elements are chosen to calculate the minimum energy state during iterative calculation, and the coordinates of each atom in the system can be obtained.

Figure 3 shows the interaction of van der Waals force of atoms A and atoms L among atoms.

As is shown in **Figure 3**, stiffness matrix for unit atom A is thus expressed as:

$$K_A = \begin{bmatrix} \left(\frac{\partial^2 E_{total}}{\partial a_A \partial a_A}\right)_{2\times2} & \left(\frac{1}{2}\frac{\partial^2 E_{total}}{\partial a_A \partial a_i}\right)_{2\times18} \\ \left(\frac{1}{2}\frac{\partial^2 E_{total}}{\partial a_i \partial a_A}\right)_{18\times2} & 0_{18\times18} \end{bmatrix} \tag{22}$$

The right endpoint of the vector of atom A can be expressed as:

$$F_A = \begin{bmatrix} \left(\overline{F}_A - \frac{\partial E_{total}}{\partial a_A}\right)_{2\times1} \\ 0_{18\times1} \end{bmatrix} \tag{23}$$

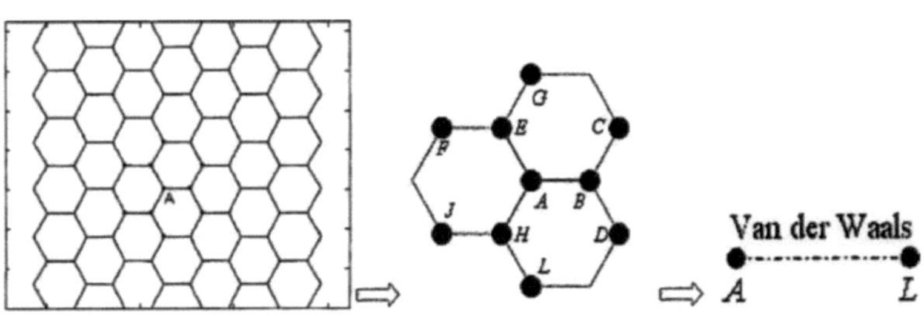

Figure 3.
Interaction of van der Waals force among atoms.

where a_A represents the coordinates of atomic vector, and in the two-dimensional case, $a_A = \begin{bmatrix} x_A & y_A \end{bmatrix}^T$; the range of i is *B-J*, and the sum of the order number of K_A and F_A is related to the interaction between atoms A and B.

For F_A, the component $\frac{\partial E_{total}}{\partial a_A} = \begin{bmatrix} \frac{\partial E_{total}}{\partial x_A} & \frac{\partial E_{total}}{\partial y_A} \end{bmatrix}^T$; For K_A, the component $\frac{\partial^2 E_{total}}{\partial a_A \partial a_A}$.

For component F_A:

$$\frac{\partial U(r_{AL})}{\partial x_A} = \frac{\partial U(r_{AL})}{\partial r_{AL}} \frac{\partial r_{AL}}{\partial x_A} \tag{24}$$

$$\frac{\partial U(r_{AL})}{\partial y_A} = \frac{\partial U(r_{AL})}{\partial r_{AL}} \frac{\partial r_{AL}}{\partial y_A} \tag{25}$$

For component K_A:

$$\frac{\partial U^2(r_{AL})}{\partial x_A \partial x_A} = \frac{\partial^2 U(r_{AL})}{\partial r_{AL}^2} \left(\frac{\partial r_{AL}}{\partial x_A}\right)^2 + \frac{\partial U(r_{AL})}{\partial r_{AL}} \frac{\partial^2 r_{AL}}{\partial x_A^2} \tag{26}$$

$$\frac{\partial U^2(r_{AL})}{\partial y_A \partial y_A} = \frac{\partial^2 U(r_{AL})}{\partial r_{AL}^2} \left(\frac{\partial r_{AL}}{\partial y_A}\right)^2 + \frac{\partial U(r_{AL})}{\partial r_{AL}} \frac{\partial^2 r_{AL}}{\partial y_A^2} \tag{27}$$

$$\frac{\partial U^2(r_{AL})}{\partial x_A \partial y_A} = \frac{\partial U^2(r_{AL})}{\partial y_A \partial x_A} = \frac{\partial^2 U(r_{AL})}{\partial r_{AL}^2} \frac{\partial r_{AL}}{\partial x_A} \frac{\partial r_{AL}}{\partial y_A} + \frac{\partial U(r_{AL})}{\partial r_{AL}} \frac{\partial^2 r_{AL}}{\partial x_A \partial y_A} \tag{28}$$

As is known to us, $U^{VdW}(r) = 4\varepsilon\left[\left(\frac{\sigma}{r_{ij}}\right)^{12} - \left(\frac{\sigma}{r_{ij}}\right)^{6}\right]$, so:

$$\frac{\partial U(r_{AL})}{\partial r_{AL}} = 4\varepsilon\left[-12 \times \frac{\sigma^{12}}{r_{AL}^{13}} + 6 \times \frac{\sigma^6}{r_{AL}^7}\right] \tag{29}$$

$$\frac{\partial^2 U(r_{AL})}{\partial r_{AL}^2} = 4\varepsilon\left[156 \times \frac{\sigma^{12}}{r_{AL}^{14}} - 42 \times \frac{\sigma^6}{r_{AL}^8}\right] \tag{30}$$

$$\frac{\partial r_{AL}}{\partial x_A} = \frac{x_A - x_L}{r_{AL}} \tag{31}$$

$$\frac{\partial^2 r_{AL}}{\partial x_A^2} = \frac{1}{r_{AL}} - \frac{x_A - x_L}{r_{AL}^2} \frac{\partial r_{AL}}{\partial x_A} \tag{32}$$

$$\frac{\partial r_{AL}}{\partial y_A} = \frac{y_A - y_L}{r_{AL}} \tag{33}$$

$$\frac{\partial^2 r_{AL}}{\partial y_A^2} = \frac{1}{r_{AL}} - \frac{y_A - y_L}{r_{AL}^2} \frac{\partial r_{AL}}{\partial y_A} \tag{34}$$

$$\frac{\partial^2 r_{AL}}{\partial x_A \partial y_A} = \frac{\partial^2 r_{AL}}{\partial y_A \partial x_A} = -\frac{(x_A - x_L)(y_A - y_L)}{r_{AL}^2} \tag{35}$$

In summary, there are two types of interactions, including bonded and nonbonded: (1) for bonded type, it usually corresponds to strong covalent bonds, the number of neighbors is preset and is dictated by the lattice. Take the CNT for example, the carbon covalent bond often has carbon atoms 3–4 neighbors, and preset list of neighbors for each atom may be used throughout the calculations, used for crystalline solids with well-defined bonds. Typical structure example is graphene; (2) for nonbonded type, it commonly corresponds to relatively weak forces, and the number of neighbors may be changing and is found based on the cutoff radius. It can be used to represent van der Waals and Coulomb forces of solid/crystal unit.

Background of atomic finite element method (AFEM) contains the C—C covalent bonds [8] by beam element and the interatomic bonds modeled [22] by nonlinear spring elements of AFEM, shown in **Figure 4**.

From **Figure 4**, the ionic bond or metallic bond is taken into consideration, van der Waals interaction and electrostatic interaction (Coulomb force) can be separately calculated. Besides, modeling of covalent bond is based on the theory of molecular mechanics method (MM). AFEM can be readily linked with the conventional continuum FEM since they are in the same theoretical framework. Based on the AFEM modeling, the elastic stiffness tensors can be described and defined [28].

2.2.2 Description of AFEM modeling hypotheses

The expression of the total steric potential energy U is given out [23]. For ionic crystal, the Coulomb force should also be considered. So the ideal expression of U, considering the electrostatic interaction, is a sum of energies due to valence or bonded interactions and nonbonded interactions:

$$U = \sum U_r + \sum U_\theta + \sum U_\phi + \sum U_\omega + \sum U_{vdw} + \sum U_{coul} \tag{36}$$

where U_r is for a bond stretch interaction, U_θ for a bond angle bending, U_ϕ for dihedral angle torsion, U_ω for an improper (out-of-plane) torsion, U_{vdw} for a nonbonded van der Waals interaction, and U_{coul} for Coulombic (electrostatic) interactions. Equivalent beam elements [8] can be used to represent covalent bond to calculate U_r, U_θ, U_ϕ, and U_ω of Eq. (36), where the molecules' chemical energy is equivalent to the beam strain energy. The coupled field of van der Waals force and Coulomb force between atoms or charged ions may be obtained by using Lennard-Jones potential or other potentials.

Nonlinear spring elements can represent U_{vdw} and U_{coul} of ionic bond. Around the equilibrium position, van der Waals force is developed by the first-order Taylor:

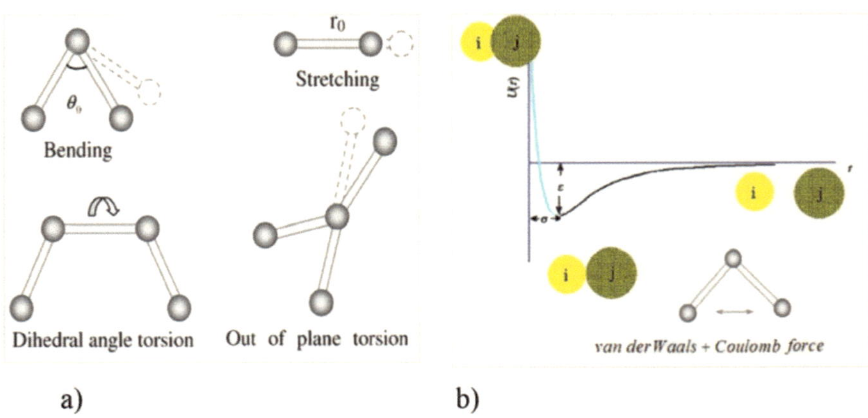

Figure 4.
AFEM modeling hypotheses of covalent bond and ionic bond. (a) Covalent bond. (b) Ionic bond or metallic bond.

$$F^{VdW}(r_i) = F^{VdW}(r_0) + \frac{dF(r_0)}{dr_i}(r_i - r_0) = 25\frac{\varepsilon}{\sigma}\left[2\left(\frac{\sigma}{r_0}\right)^{13} - \left(\frac{\sigma}{r_0}\right)^{7}\right] - 24\frac{\varepsilon}{\sigma}\left[26\left(\frac{\sigma}{r_0}\right)^{14} - 7\left(\frac{\sigma}{r_0}\right)^{8}\right](r_i - r_0) \tag{37}$$

where r_0 (unit: Å) is the distance between two atoms around the equilibrium position without external force, r_i is the distance of the Lennard-Jones potential interaction, ε is the bond energy of van der Waals interaction between two balanced atoms, and σ is the value of r at which the potential is zero [29].

For arbitrary atom type of A-B, the van der Waals force can be obtained as follows:

$$F_{VdW}^{atomA-atomB}(r) = 24\frac{\varepsilon^{A-B}}{\sigma^{A-B}}\left[2\left(\frac{\sigma^{A-B}}{r^{A-B} - r_0^{A-B}}\right)^{13} - \left(\frac{\sigma^{A-B}}{r^{A-B} - r_0^{A-B}}\right)^{7}\right] - F_{VdW}^{atomA-atomB}\left(r_0^{A-B}\right) \tag{38}$$

For U_{vdw}, the nonlinear displacement-force characteristic of spring element as the first derivative of potentials must be defined by means of the program potentials [22], then by the coordination transformation.

Coulomb potential of ionic crystal according to Coulomb's law is given by:

$$U_{ij}^{Coulomb} = q_i q_j (r_i - r_j) / \left(4\pi\varepsilon_0 \cdot |r_i - r_j|^2\right) \tag{39}$$

where q_i and q_j are partial charges, ε_0 is the dielectric permittivity of vacuum (8.85419×10^{-12} F/m).

Nonlinear spring element is used to represent interatomic force of Lennard-Jones potential. For U_{vdw} and U_{coul} represented by nonlinear spring element, since transforming from $[x, y, 1]$ to $[x', , y', 1]$ by the matrix $[T]$, the expression is as:

$$[x', , y', 1] = [x, y, 1] \cdot T = [x, y, 1] \cdot \begin{bmatrix} 1, 0, 0 \\ 0, 1, 0 \\ r_0, F_0, 0 \end{bmatrix} = [x + r_0, y + F_0, 1] \tag{40}$$

where r_0 and F_0 are the offset values in the x- and y-directions, respectively.

2.2.3 Interactions between atoms by Lennard-Jones potential

Generally, the expression of van der Waals force can be obtained by the Lennard-Jones potential [30]:

$$U_{ij}^{VdW}(r) = U_{repuls\ if} + U_{attract\ if} = \sum_i \sum_j 4\varepsilon_{ij}\left[\left(\frac{\sigma_{ij}}{r_{ij}}\right)^{12} - \left(\frac{\sigma_{ij}}{r_{ij}}\right)^{6}\right] \tag{41}$$

where r_{ij} is the distance between two atoms i and j; ε_{ij} and σ_{ij} are parameters of Lennard-Jones. ε_{ij} represents the minimum potential energy between atoms i and j, which corresponds to the most stable interaction (the depth of the potential well under the distance of σ_{ij}). σ_{ij} is the equilibrium distance where atomic energy between two atoms is zero.

Schematic illustration of Lennard-Jones potential and van der Waals force is shown in **Figure 5**.

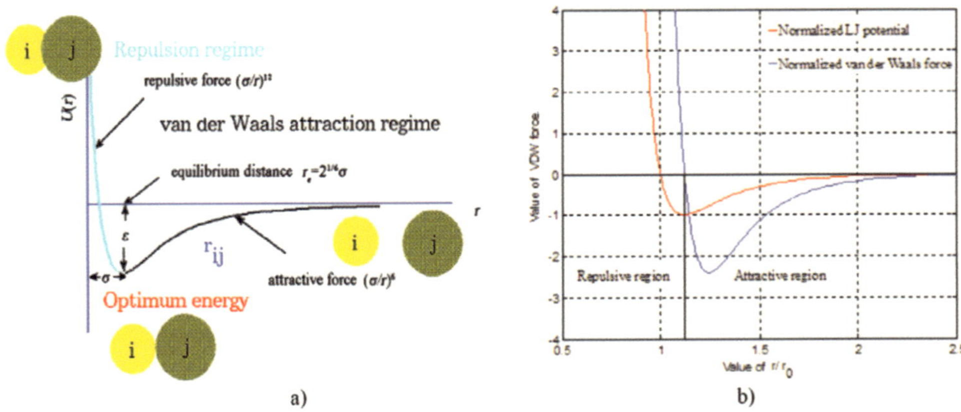

Figure 5.
Schematic illustration of Lennard-Jones potential and van der Waals force. (a) Schematic of Lennard-Jones potential. (b) van der Waals force of C—C bond.

As shown in **Figure 5(a)**, repulsive force is more intense but the range is very short that prevents exchange of electrons from occupying the same region of space, since the forces exerted at larger distances are very small and can be neglected [23, 31]. Besides, from **Figure 5(b)**, the equilibrium position of Lennard-Jones potential is at the place of $r/\sigma = 1$. The potential must be truncated at some point named the cutoff radius, R_{cut}, usually set to $R_{cut} = 2.5\sigma$. When the atomic distance is greater than 2.5σ, the Lennard-Jones potential is approximate to zero. As is shown in **Figure 5(b)**, if the average distance r between the atoms increases, the attractive force of van der Waals force between pairs of atoms will resist the force caused by the increase of distance r. And if the average distance r between the atoms increases, the repulsive force between pairs of atoms will be able to resist the force caused by the decrease of distance r.

Interactions of van der Waals consist of two terms: a repulsive term and an attractive term. The function of Lennard-Jones is known as a function of the potential of van der Waals type. It is used to describe the intermolecular force between argon atoms i and j separated by a distance r_{ij}. Any two molecules attract each other at long separation distance and repel each other when they come closer [32].

Moreover, coordinate transformation is used, and these curves are translated so that the system at the beginning of the calculation is at the equilibrium (no more attractive/repulsive force). When the interatomic spacing is greater than its unstressed value, the attractive forces between atoms must be greater than the repulsive forces (the attractive forces balance both the repulsive forces and the external forces). Repulsive forces are taken as positive and attractive as negative. This conveniently makes their potential energies positive and negative, respectively, in which the zero of potential energy is taken at large separation. Thus, AFEM can be used in multiscale computation with applications to carbon nanotubes and CH [28, 33].

2.2.4 Modeling of ionic crystal based on AFEM method

AFEM modeling of zigzag CNT and zigzag graphene thus can be described by the algorithm, in **Figure 6**.

As is shown in **Figure 6**, covalent bonds (e.g., for C—C in the studied structures) are characterized by an equivalent beam, and nonbonded potential is characterized by using nonlinear spring elements to represent the coupled field of van der Waals force. Then, axial moduli of the structure, as a function of elastic

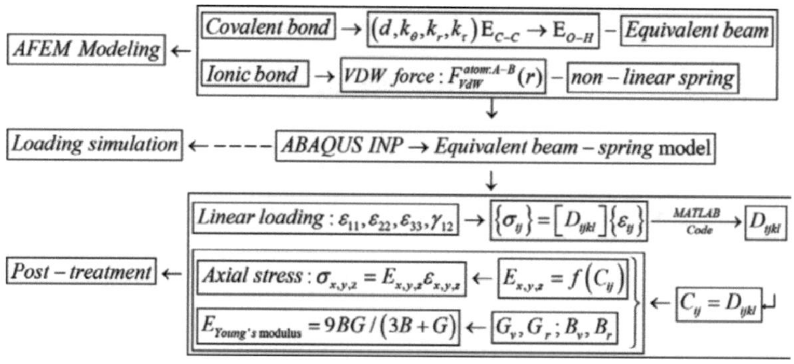

Figure 6.
Modeling of hexagonal structure based on AFEM and solving process.

constants, are derived and verified by cubic linear elements volume unit on ABAQUS. Thus, elastic moduli of TLGSs and CH hexagonal structures can be determined.

In the AFEM modeling, O—H bond and C—C bond as covalent bonds are characterized to be equivalent beam based on MM method, and nonlinear spring element is used to characterize the coupled field of van der Waals force and Coulomb force for ionic or metallic bond by using Lennard-Jones potential or Buckingham potential, which is calculated by MATLAB and then imported into input program. Moreover, axial moduli of triple-layer graphene sheet (TLGSs) and CH hexagonal structure based on AFEM as a function of elastic constants are derivated and verified by FE simulation on ABAQUS. Elasticity coefficient matrix of TLGSs and CH crystals are solved reversely, thus elastic properties of TLGSs and CH polycrystals are furtherly investigated by the well-known Reuss-Voigt-Hill estimation.

2.3 Elastic modulus determination of hexagonal crystal structures

ABAQUS as a powerful nonlinear software is used to the continuation of quantum mechanics and an approximate equivalent method—AFEM. Nonlinear springs are introduced to describe the relationship between atoms of different initial distances. When the van der Waals force is considered, the coordinate transformation needs to be done before carrying out the INP file. Input file usage:

```
*SPRING, NONLINEAR, DEPENDENCIES = n
first data line
force, relative displacement, field variable 1, etc.
...
```

The nonlinear spring behavior is not supported in ABAQUS/CAE when you define springs as engineering features. Instead, we can define connectors that have spring-like elastic behavior. To obtain the homogenized moduli, the notion of homogenization has been introduced using the representative volume element (RVE) of the material [34]. When a solid model is under external load, microstresses and strains are induced; according to the statistical homogeneity assumption, an appropriate RVE can be defined and isolated. On the RVE boundary, there exist definitive surface displacements and surface tractions. Within the RVE, there exist definitive stress field σ_{ij} and strain field ε_{ij}. Through homogenization [35], mechanical behavior is described by a definitive constitutive law. So, the mechanical properties of anisotropic solids can be investigated using FEM simulation by RVE unit on ABAQUS.

2.3.1 Axial moduli determination of hexagonal structures

Stress-strain relation in an orthotropic hexagonal crystal can be defined by the independent elastic stiffness parameters of ABAQUS manual [36]:

$$\begin{bmatrix}\sigma_{11}\\ \sigma_{22}\\ \sigma_{33}\\ \sigma_{12}\\ \sigma_{13}\\ \sigma_{23}\end{bmatrix}=\begin{bmatrix}c_{11} & c_{12} & c_{13} & 0 & 0 & 0\\ & c_{11} & c_{13} & 0 & 0 & 0\\ & & c_{33} & 0 & 0 & 0\\ & \text{sym} & & c_{44} & 0 & 0\\ & & & & c_{44} & 0\\ & & & & & 1/2\times(c_{11}-c_{12})\end{bmatrix}\begin{bmatrix}\varepsilon_{11}\\ \varepsilon_{22}\\ \varepsilon_{33}\\ \gamma_{12}\\ \gamma_{13}\\ \gamma_{23}\end{bmatrix}$$
$$=\begin{bmatrix}D_{1111} & D_{1122} & D_{1133} & 0 & 0 & 0\\ & D_{2222} & D_{2233} & 0 & 0 & 0\\ & & D_{3333} & 0 & 0 & 0\\ & \text{sym} & & D_{1212} & 0 & 0\\ & & & & D_{1313} & 0\\ & & & & & D_{2323}\end{bmatrix}\begin{bmatrix}\varepsilon_{11}\\ \varepsilon_{22}\\ \varepsilon_{33}\\ \gamma_{12}\\ \gamma_{13}\\ \gamma_{23}\end{bmatrix} \quad (42)$$

where σ represents the normal stress and shear stress in each direction (unit: nN/nm^2); ε and γ are the normal strain and shear strain in each direction. The formula developed by finite element technique to solve the elastic constants by volume unit has been described [28], as in a manual of theory details [37].

If a unidirectional force is applied to obtain the strain value, the elastic constants of the matrix can be obtained by simultaneous equations. **Figure** 7 shows schematic deformation component.

A finite element method for solving the elastic coefficient matrix C_{ij} is proposed according to the schematic volume unit and stress direction after strain-loading (**Figure** 7**(a)**) and the different components of the deformation through different loading modes (**Figure** 7**(b)**). Supposing $A=\varepsilon_{11}, B=\varepsilon_{22}, C=\varepsilon_{33}, D=\varepsilon_{31}, E=\varepsilon_{11}+\varepsilon_{21}$ (here ε_{ij} independent strain of each unit in all directions), as elastic constants C_{ij} = D^{el}, Eq. (42) can be changed as:

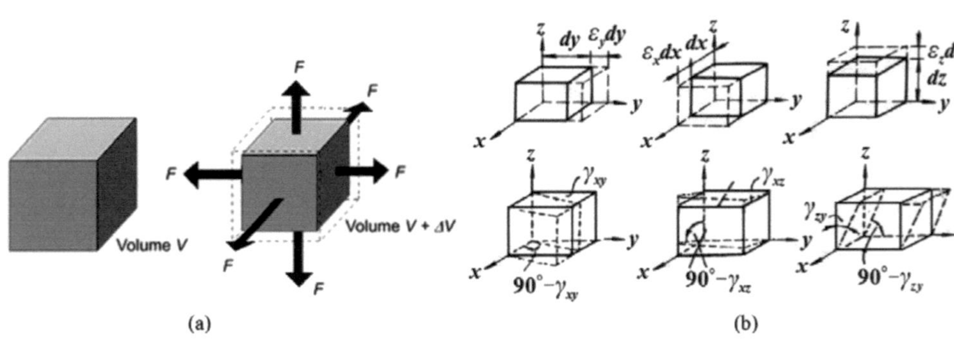

Figure 7.
Schematic deformation component and loading modes of linear/shear strain. (a) Volume unit and stress direction under unidirectional stress. (b) Loading modes of linear/shear strain.

$$\begin{pmatrix} A & B & C & 0 & 0 \\ B & A & C & 0 & 0 \\ 0 & 0 & A+B & C & 0 \\ 0 & 0 & E\ D & 0 & \\ 0 & 0 & 0 & 0 & \gamma_{12} \end{pmatrix} \begin{pmatrix} c_{11} \\ c_{12} \\ c_{13} \\ c_{33} \\ c_{44} \end{pmatrix} = \begin{pmatrix} \sigma_{11} \\ \sigma_{22} \\ \sigma_{33} \\ 0 \\ \sigma_{12} \end{pmatrix}$$

$$\rightarrow \begin{cases} c_{11} = D_{1111} = D_{2222} = \dfrac{A\sigma_{11} - B\sigma_{22}}{A^2 - B^2} - \dfrac{CD\sigma_{33}}{D(A+B)^2 - C(A+B)E} \\ c_{12} = D_{1122} = \dfrac{A\sigma_{22} - B\sigma_{11}}{A^2 - B^2} - \dfrac{CD\sigma_{33}}{D(A+B)^2 - C(A+B)E} \\ c_{13} = D_{1133} = D_{2233} = \dfrac{D\sigma_{33}}{D(A+B) - CE}; \\ c_{33} = D_{3333} = -\dfrac{E\sigma_{33}}{D(A+B) - CE}; \\ c_{44} = D_{1212} = D_{1313} = \dfrac{\sigma_{12}}{\gamma_{12}}; c_{66} = \dfrac{1}{2}(c_{11} - c_{12}) = D_{2323} \end{cases} \tag{43}$$

Stress-strain relation of hexagonal crystal is presented in Eq. (42). Assuming a normal stress σ_3 in z-axis direction is introduced, it will not result in shear strains, thus: $\gamma_{12} = \gamma_{23} = \gamma_{31} = 0$. We assume σ_{33} expressed by ε_{33} as: $\sigma_{33} = E_3\varepsilon_{33}$. Substituting this equation into the ternary equations of stress-strain relation Eq. (42), if ε_{13}, ε_{23}, ε_{33} are not simultaneously zero, the determinant factor must equal to zero according to the theory of linear algebra, so homogeneous equations can be changed:

$$\begin{cases} 0 = c_{11}\varepsilon_{13} + c_{12}\varepsilon_{23} + c_{13}\varepsilon_{33} \\ 0 = c_{12}\varepsilon_{12} + c_{11}\varepsilon_{23} + c_{13}\varepsilon_{33} \\ 0 = c_{13}\varepsilon_{13} + c_{13}\varepsilon_{23} + (c_{33} - E_3)\varepsilon_{33} \end{cases} \rightarrow \begin{vmatrix} c_{11} & c_{12} & c_{13} \\ c_{12} & c_{11} & c_{13} \\ c_{13} & c_{13} & c_{33} - E_3 \end{vmatrix} = 0 \tag{44}$$

Eq. (44) can be simplified as:

$$\begin{vmatrix} c_{11} & c_{12} & c_{13} \\ c_{12} & c_{11} & c_{13} \\ c_{13} & c_{13} & c_{33} - E_3 \end{vmatrix} \underline{\underline{R_2 - R_1}} \begin{vmatrix} c_{11} & c_{12} & c_{13} \\ c_{12} - c_{11} & c_{11} - c_{12} & 0 \\ c_{13} & c_{13} & c_{33 - E_3} \end{vmatrix} \underline{\underline{C_2 + C_1}} \begin{vmatrix} c_{11} & c_{11} + c_{12} & c_{13} \\ c_{12} - c_{11} & 0 & 0 \\ c_{13} & 2c_{13} & c_{33} - E_3 \end{vmatrix}$$

$$= \begin{vmatrix} c_{11} + c_{12} & c_{13} \\ 2c_{13} & c_{33 - E_3} \end{vmatrix} = 0.$$

Then, we can get the relationship between z-axial modulus E_3 and elastic constants C_{ij}, the formula of *z*-axial modulus E_3 is thus obtained:

$$E_3 = c_{33} - \frac{2c_{13}^2}{c_{11} + c_{12}} \tag{45}$$

Similarly, we can obtain relationship between *y*-axial modulus E_2 and elastic constants as:

$$E_2 = (c_{11} - c_{12})\left(1 + \frac{c_{12}c_{33} - c_{13}^2}{c_{11}c_{33} - c_{13}^2}\right) \tag{46}$$

Finally, we can see that formulas of E_1 and E_2 axial moduli in x- and y-directions are the same, so:

$$E_1 = E_2 = (c_{11} - c_{12})\left[1 + (c_{12}c_{33} - c_{13}^2)/(c_{11}c_{33} - c_{13}^2)\right] \quad (47)$$

Based on elastic constants calculated, homogenized elastic properties of polycrystals can be calculated. Firstly, shear and bulk moduli are obtained by means of Voigt and Reuss bounds, which are based on the c_{ij} calculated by Wu et al. [38]. Then, Young's modulus can be finally determined from shear/bulk modulus formulas.

2.3.2 Elastic coefficient solution by finite element method

As $\varepsilon_1 = \varepsilon_a$, $\varepsilon_3 = \varepsilon_c$, so ε_b can be easily obtained by the relative displacement in b-direction.

$$\vec{b}\begin{pmatrix} -\cos 60^\circ \\ \sin 60^\circ \\ 0 \end{pmatrix} = \vec{b}\begin{pmatrix} -1/2 \\ \sqrt{3}/2 \\ 0 \end{pmatrix}, \quad \varepsilon_b = \begin{pmatrix} -1/2 \\ \sqrt{3}/2 \\ 0 \end{pmatrix} \cdot \begin{pmatrix} c_{11} & c_{12} & c_{13} \\ c_{12} & c_{11} & c_{13} \\ c_{13} & c_{13} & c_{33} \end{pmatrix} \cdot \begin{pmatrix} -1/2 \\ \sqrt{3}/2 \\ 0 \end{pmatrix} \quad (48)$$

For hexagonal structure, the coordinate transformation from 90 to 120° (in **Figure 8**), the relation is as:

Under uniaxial condition ($\varepsilon_{12} = \varepsilon_{13} = \varepsilon_{23} = 0$), so

$$\varepsilon_b = \frac{1}{4}\varepsilon_{11} + \frac{3}{4}\varepsilon_{22}, \quad \varepsilon_{22} = \frac{4}{3}\varepsilon_b - \frac{1}{3}\varepsilon_{11} \quad (49)$$

Linear elasticity in an orthotropic material can also be defined by giving the independent elastic stiffness parameters, as functions of predefined fields. ABAQUS/standard spatially varying orthotropic elastic behavior can be defined for homogeneous solid continuum elements by using a distribution. The distribution must include default values for the elastic moduli. If a distribution is used, no dependencies on temperature and/or field variables for the elastic constants can be defined and determined on ABAQUS.

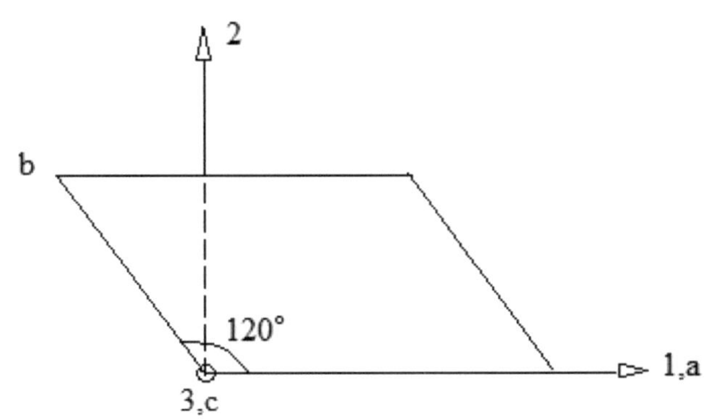

Figure 8.
Solving equivalent strain conversion from 90 to 120°.

3. Application to structures of zigzag CNT and zigzag SLGSs

For application to structures of CNT and graphene by AFEM, a calculation formula of Young's modulus of zigzag SWCNT and zigzag SLGSs is given out and verified. As the total potential energy is equal to the thin cylindrical shell strain energy, the elastic modulus of $(n, 0)$ zigzag nanotube is developed by reference [13] without considering the thickness as:

$$E_{CNT}^{zigzag} = \frac{4\sqrt{3}KC}{9C + 4Kl_0^2\left(\lambda_{z1}^2 + 2\lambda_{z2}^2\right)} \tag{50}$$

where K, C, respectively, correspond to constants of bond angle changes and bond stretching; λ_{z1} and λ_{z2} are structure parameters of zigzag nanotube, $\lambda_{z1} = -3\sqrt{4 - 3\cos^2\left(\frac{\pi}{2n}\right)} \cdot \cos\left(\frac{\pi}{2n}\right) / \left[8\sqrt{3} - 2\sqrt{3}\cos^2\left(\frac{\pi}{2n}\right)\right]$, and $\lambda_{z2} = 12 - 9\cos^2\left(\frac{\pi}{2n}\right) / \left[16\sqrt{3} - 4\sqrt{3}\cos^2\left(\frac{\pi}{2n}\right)\right]$. Young's modulus formulas of the graphene sheet can be derived when n becomes ∞ in zigzag nanotube parameters λ_{z1} and λ_{z2} expression, which is as follows:

$$E_{SLGSs}^{zigzag} = \frac{8\sqrt{3}KC}{18C + Kl_0^2} \tag{51}$$

where K, C, respectively, correspond to constants of bond angle changes and bond stretching. So, the Young's modules of C—C covalent bonds of zigzag SWCNT and SLGSs can be verified by reference [13].

Axial direction restraint and transverse direction with freedom (to ensure axial cross section during tensile can be freely deformed with shrinkage), the concentrated tensile force is added on each node of the other end. The sum of axial centralized force vectors at each node of carbon nanotubes is added up to F, and the finite element calculation is continuously carried out on ABAQUS. Young's modulus is obtained as:

$$E = \frac{\sigma}{\varepsilon} = \frac{F/A}{\Delta l/l} = \frac{F \cdot l}{\pi D \cdot t \cdot \Delta l} \tag{52}$$

where A is the cross-sectional area, $A = \pi Dt$; D is the carbon nanotube diameter, and t is the wall thickness. The boundary conditions are free-fixed.

3.1 Nanoscale modeling of zigzag SWCNT and SLGSs

3.1.1 Definition of C—C bond by coordination transformation

The transformed van der Waals curves are calculated, and the one-spring and two-spring of VDW forces are tested on ABAQUS software, as in **Figure 9**.

From **Figure 9(a)** and **(b)**, the definition of van der Waals forces and the coordinate conversion of C—C bond are tested on ABAQUS, spring step-election x is 0.001 nm in order to meet the accuracy requirements. For C—C bond, parameters are as: r_{max} = 0.425 nm, F_{max} = 7.0174 nN. When the distance r = 0.3833208 nm, the van der Waals force is zero, whereas the equilibrium distance r_0 is 0.3415 nm, and the F_0 is −3.9247 nN.

3.1.2 Modeling of zigzag CNT and zigzag SLGSs

Modeling of a finite number graphene and (10, 0) zigzag CNT are shown in **Figure 10**.

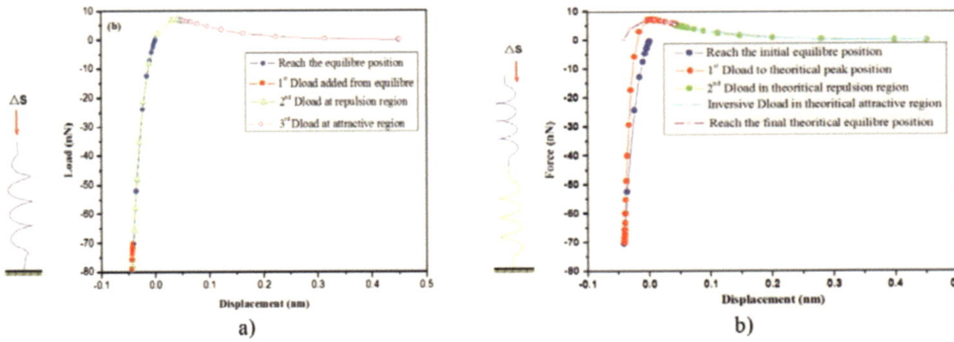

Figure 9.
One-spring and two-spring tests of transformed van der Waals curves. (a) True VDW of one-spring test. (b) True VDW of two-spring test.

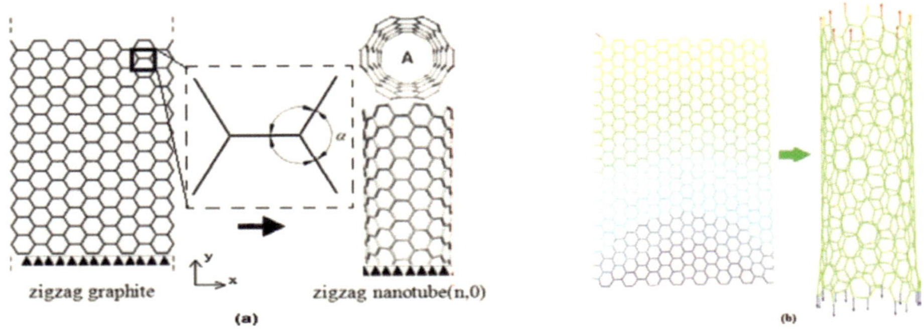

Figure 10.
AFEM modeling of zigzag-CNT and single-layer graphene sheet (SLGSs). (a) Structures of zigzag graphene and SWNCT. (b) Models of zigzag graphene and SWNCT.

The boundary conditions of SLGSs and zigzag CNT with the fixed bound at x-direction and the tensional force applied at y-direction are shown in **Figure 10(a)**. Moreover, the size *3.339* nm × *4.446* nm of SLGSs and (10, 0) zigzag CNT modeled with ABAQUS are shown in **Figure 10(b)** to apply the tensional force in y-direction. Parameters of C—C covalent bonds used in simulation are as [13, 39]: C—C bond length is 0.142 nm, thickness is 0.3414 nm, bond angle θ_0 is 120°, balanced distance constant $\sigma_{c\text{-}c}$ is 0.3414 nm, and bond energy constant $\varepsilon_{c\text{-}c}$ is 0.00239 eV. After we obtain the reaction force of y-direction, we substitute all parameters into the formula of E_{SLGSs} and E_{CNT} in Eq. (52), the results are as: E_{SLGSs} = 0.867 TPa, E_{CNT} = 0.847 TPa.

3.2 Young's moduli of zigzag SWCNT and SLGSs

Here, we choose K = 742 nN/nm and C = 1.42 nN.nm; When n becomes ∞ (here $\lambda_{z1} = -0.28867$; $\lambda_{z2} = 0.14433$), Young's modulus of graphene by Eq. (51) is calculated. **Figure 11** shows the Young's modulus variation of zigzag SWCNT with various diameters and the thickness and Young's modulus relation.

Figure 11(a) provides the result of the present modeling of zigzag SWCNT and SLGSs with a thickness of 0.34 nm, and the comparison of present results (E_{SLGSs} = 1.06 TPa) and literature results (E_{CNT} = 1.05 TPa [40]) for various thicknesses are also added. It can be seen in **Figure 11(b)** that the result of $E.t$ (product of the Young's modulus and the thickness) tends to the analytical value found by molecular mechanics approach [13]. Young's modulus variation of zigzag nanotubes with the diameter is shown in **Figure 11**, the result is in good agreement

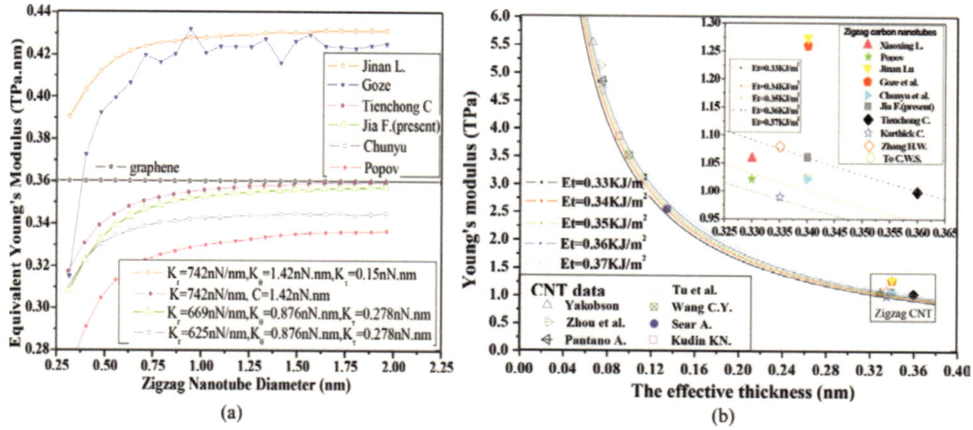

Figure 11.
Young's modulus variation and the thickness and Young's modulus relation of zigzag SWCNT. (a) Young's modulus variation with various diameters. (b) The thickness and Young's modulus relation.

with zigzag nanotubes [8–10, 13, 14, 20, 21] and no-distinguish CNT [5, 11, 12, 15–18]. When the thickness of carbon nanotubes chooses to be the experience value 0.34 nm, the elasticity modulus E of carbon nanotubes calculated tends to 1.05 TPa, which is close to the consistent results of Yakobson et al. [5] and the result of the force-constant models by Popov et al. [10] and an analytical model based on a molecular mechanics approach by Chang and Gao [13]. The main difference with references [8, 13, 20] stems from the different values of the chemical bond parameters $\{k_r, k_\theta, k_\tau\}$. It can be seen in **Figure 11(b)** that the numerical result of $E.t$ of zigzag CNT with various thicknesses is plotted, we can see that $E.t$ is mainly distributed in the range of 0.33–0.37 kJ/m^2, and our result of $E.t$ is also consistent with literature findings. It therefore shows that the atomic finite element analysis to study a covalent bond is feasible.

4. Application to the nanostructure of the triple-layer graphene

In general, the configuration of the equilibrium solid is decided by the minimum energy state. The basic idea of the continuum finite element is to divide into a certain number of units, of which the cell is represented by the nodes of the finite number of discrete. The location of these nodes is determined by the energy minimization state, which is similar to the basic idea of molecular statics to calculate the position of the atoms. The point is to minimize the energy of molecular static system for determining the position of each atom, that is why the concept of cohesion is introduced to determine the displacement of each point. For multilayered graphene, two interatomic potentials are used [41, 42]: one is for in-layer bonded interactions—multibody (Tersoff-Brenner), the other for interlayer nonbonded interactions (Lennard-Jones). A multilayer graphene modeling considering van der Waals force can be described by introducing nonlinear spring element between layers, which can be modeled by the AFEM method.

4.1 C—C bond definition and nanoscale modeling of graphene

AFEM modeling of triple-layer graphene sheet (TLGSs) is shown in **Figure 12**.
Graphene crystal is shown in **Figure 12(a)**. A round-section definition of equivalent graphene-beam is: k_r = 669 nN/(nm^2), k_θ = 0.876 nN.nm, and k_τ = 0.278 nN.

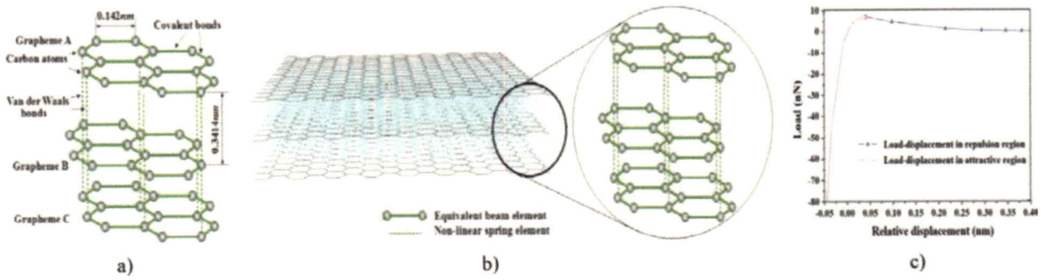

Figure 12.
AFEM modeling of triple-layer graphene sheet (TLGSs). (a) Graphene crystal. (b) TLGSs model with beam-spring element. (c) van der Waals force of C—C bond.

nm [8]. The size 3.41 × 3.46 × 0.680 nm of TLGSs is modeled with ABAQUS in **Figure 12(b)** to determine its elastic constants. van der Waals force of C—C bond is shown in **Figure 12(c)**. Parameters of Lennard-Jones potential is as [30]: $U_{vdw} = -15.2/r^6 + 24.1 \times 10^3/r^{12}, r_0 = \sqrt[6]{2}\sigma = 0.383nm$. Parameters of graphene used in simulation are as [24, 43]: C—C bond length is 0.142 nm, thickness is 0.3414 nm, bond angle θ_0 is 120°, balanced distance constant $\sigma_{c\text{-}c}$ is 0.3414 nm, and bond energy constant $\varepsilon_{c\text{-}c}$ is 0.00239 eV.

C—C bonds are represented by means of equivalent beams, for triple-layer graphene sheets (TLGSs) model, interlayer is mainly maintained by van der Waals force, thus it can be defined by Lennard-Jones potential and connected by nonlinear-spring element. The cohesive energy between layers is also considered [44]. Here the C-C bond distance σ = 0.3415 nm, spacing distance is as: $h_1 = 0.992\sigma_{C\text{-}C}$ = 0.3373 nm, $h_2\text{-}h_1 = 0.997\sigma_{C\text{-}C}$ = 0.3390 nm.

4.2 Elastic constants and homogenized moduli of graphene

Similarly, elastic constants of TLGSs are verified by volume unit. Then, we extend to the equivalent beam-spring model by AFEM. Diagram of strain loading in triple-layer graphene sheets is as shown in **Figure 13**.

Strain loading of triple-layer graphene sheets is shown in **Figure 13**. Then, the strain-stress data of TLGSs with beam element and beam-spring element separately by FEM and AFEM are given in **Table 3**.

From **Table 3**, C_{ij} is calculated and axial modulus by Eqs. (45)–(47) is as: $\sigma_x = \sigma_y = 1025.21\varepsilon_y$, $\sigma_z = 36.14\varepsilon_z$. As the graphene layer plane exists in the *X-Y* plane and the thickness direction of *Z* is relatively small, axial modulus of SLGSs can be averaged *as:* $E = (E_x + E_y)/2$ = 1025.81 GPa, with comparison of 996.49 GPa by Berryman [45]. Homogenized moduli of TLGSs based on AFEM can be calculated, and elastic moduli of polycrystalline graphene are given in **Table 4**.

From references of Berryman [46, 47], elastic moduli for polycrystalline graphenes can be calculated based on the elastic constant matrix C_{ij} of graphene. Elastic constants and elastic modulus by AFEM are in agreement with results of Berryman [48]. Stress-strain and elastic moduli of TLGSs by AFEM are shown in **Figure 14**.

From **Figure 14(a)**, elastic constants are in quite good agreement; axial modulus in Z direction is much smaller than that of X, Y direction. From **Figure 14(b)**, elastic moduli by AFEM are nevertheless slightly lower than values by Berryman. The nonspherical nature in Young's modulus of TLGSs is anisotropic and the *x*-axis is more anisotropic than the *z*-axis because of the values of C_{33} larger than C_{11}. Elastic moduli of TLGSs by Wu et al. [38] are as: B = 167.55 GPa, G = 118.21 GPa. Thus, Young's modulus of TLGSs is averaged to be 0.525 TPa, which is lower than SLGSs; this trend is consistent with the trend between the bilayer graphene sheets

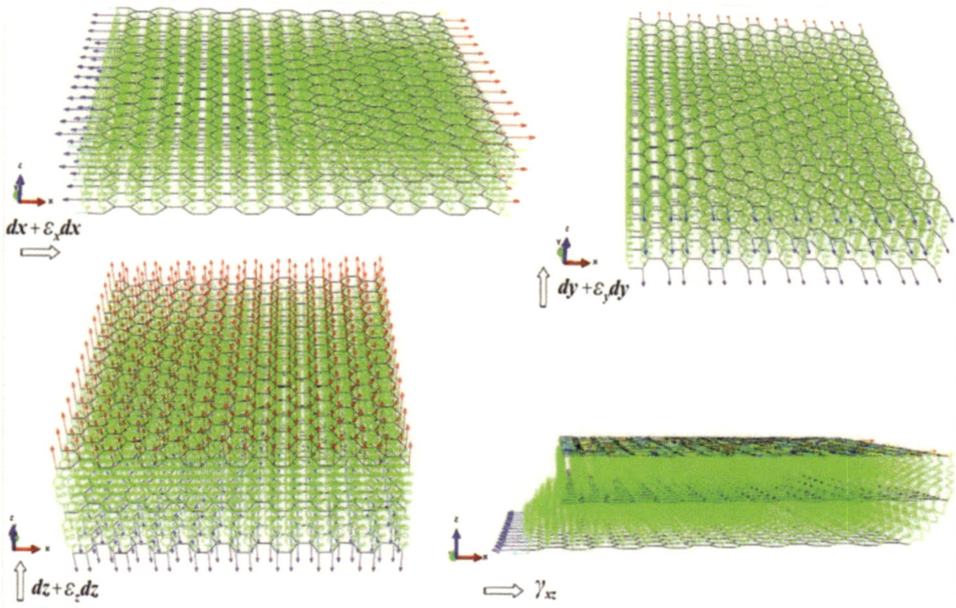

Figure 13.
Strain loading of triple-layer graphene sheets.

(BLGSs) of 0.8 TPa [49] and SLGSs of 0.996 TPa [45], showing van der Waals forces between layers has weakened the Young's modulus value of multigraphene.

5. Application to the hexagonal portlandite (CH) structure

CH is a typical nanostructure of the microscale-hydrated Portland cement paste [50] and its Young's modulus is needed in a 3D multiscale mortar model [51], so we aim to propose a modeling method based on AFEM to investigate anisotropic stiffness of CH crystalline and elastic properties of polycrystals [52].

5.1 Nanoscale modeling and coupling force field of CH

5.1.1 Nanoscale modeling of CH ionic crystal

CH is a hexagonal structure; its crystal lattice [53] is: $a = b = 3.5930$ Å, $c = 4.9090$ Å, $\alpha = \beta = 90°$, $\gamma = 120°$. AFEM modeling of portlandite (CH) is shown in **Figure 15**.

From **Figure 15(a)**, the Ca^{2+} is octahedrally coordinated by oxygen. There is no hydrogen bond across the layer. The direction of hydroxyl groups formed by oxygen and hydrogen is vertical to the (001) plane direction. The atomic distance between O^{2-} and H^{+} is 2.757 Å. O—H covalent bond can be solved by a linear bond energy ratio relative to C—C bond energy. For O—H covalent bond, $L = 0.985$ nm [54]. Portlandite is modeled with equivalent beam-spring element in **Figure 15(b)**. The potential cutoff radius is 0.8 nm.

Parameters of weight, charges, and Lennard-Jones potential are given in **Table 5**.

5.1.2 Coupling force field definition of CH ionic crystal

For CH modeling, the initial conditions are as: (1) no hydrogen bond or other strong bonds across the layer; (2) without considering the effect of three-body potential and

Strain-stress by volume unit anisotropy				Strain-stress by beam-spring element			
σ_{ij} (nN/nm^2)	ε_{11}	ε_{22}	ε_{33}	σ_{ij} (nN/nm^2)	ε_{11}	ε_{22}	ε_{33}
σ_{11} = 902.32	0.879739	−0.145212	−0.301956	σ_{11} = 957.769	0.8797	−0.15202	−0.112044
σ_{22} = 889.883	−0.143092	0.8676238	−0.297795	σ_{22} = 944.594	−0.14993	0.86762	−0.112044
σ_{33} = 31.8732	−0.0106694	−0.0106694	0.882005	σ_{33} = 15.641	−5.278e-3	−5.403e-3	0.441003
σ_{12} = 3.0591	γ_{12} = 0.6294			σ_{12} = 2.256	γ_{12} = 0.8797		

Table 3.
Strain-stress data of grapheme by volume unit using FEM and beam-spring element using AFEM modeling.

Parameters	C_{11}	C_{12}	C_{13}	C_{33}	C_{44}	C_{66}	*Bv*	*Br*	*Gv*	*Gr*	*E*	μ
FEM	1057.21	183.49	15.00	36.49	4.86	436.84	286.43	35.75	218.48	11.00	507.27	0.212
AFEM	1123.02	201.33	5.50	35.72	2.56	460.82	300.71	34.39	231.15	6.07	525.01	0.214
Equation-AFEM	1123.10	201.39	5.12	38.24	2.61	460.86	300.86	36.65	231.40	6.20	526.39	0.215
Equation-FEM	1057.51	183.83	13.29	45.24	4.86	436.83	286.78	43.64	219.30	11.19	511.53	0.217
Graphene [45]	1060.00	180.00	15.00	36.50	4.5	440.00	286.28	35.76	219.57	10.26	507.86	0.212

Table 4.
Elastic constants of grapheme and homogenized moduli (bulk, shear, and Young's moduli) of polycrystalline grapheme by FEM and AFEM (unit: GPa).

DOI: http://dx.doi.org/10.5772/intechopen.84597

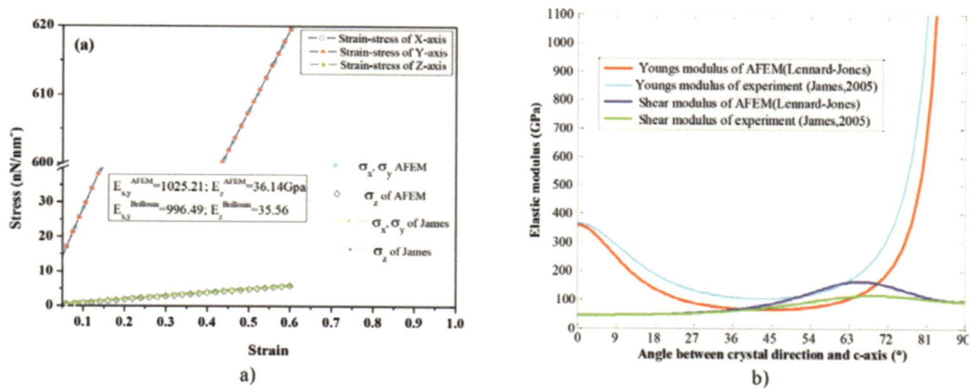

Figure 14.
Comparison of TLGSs stress-strain and elastic moduli by AFEM modeling. (a) Stress-strain and axial modulus by AFEM. (b) Comparison of Young's and shear moduli.

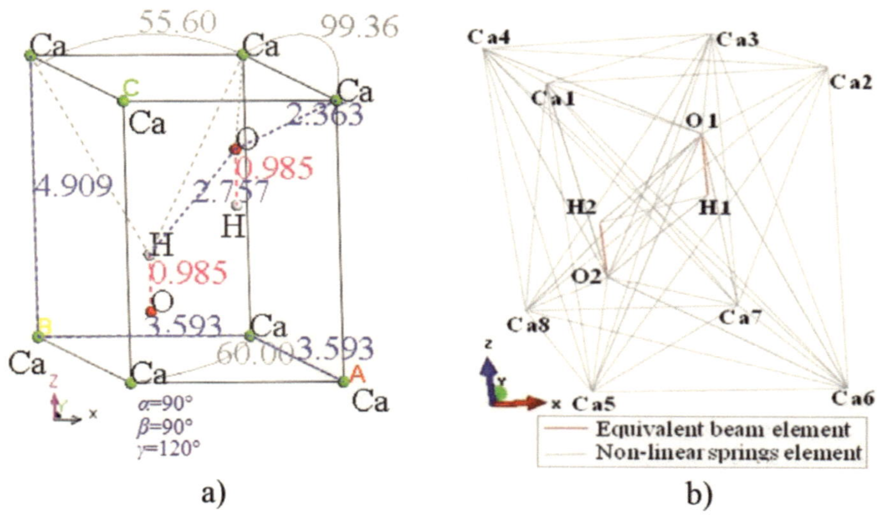

Figure 15.
AFEM modeling of portlandite (CH). (a) Portlandite crystal. (b) CH model with beam-spring element.

dihedral; (3) O—H is approximately defined by estimation of C—C bond according to the linear bond energy ratio. Moreover, CH parameters by AFEM modeling are as: O—H bond length $L_{O—H}$ = 0.986 nm, potential cutoff radius is 0.8 nm, $E_{O—H}$ = 4.611 × 10^3 nN/nm^2, $d_{O—H}$ = 0.175 nm, $G_{O—H}$ = 1.921 × 10^3 nN/nm^2. Spring element to describe the interatomic forces under the equilibrium state is shown in **Figure 16**.

From **Figure 16(a)**, the curve of VDW force has been translated so that the system at the beginning of the calculation is at the equilibrium (no more attractive/repulsive force). From **Figure 16(b)**, Coulomb force curve is given, then the coupling force field of Ca—O and O—H bond is decayed rapidly with increasing distance, and Ca—Ca and O—O bonds are weakened during the modeling of CH, as shown in **Figure 16(c)**.

5.2 Elastic constants and homogenized moduli of CH

Similarly, elastic constants of CH crystal are verified by volume unit. Then, we extend to the equivalent beam-spring model by AFEM. Strain loading of CH crystal is shown in **Figure 17**.

Weight and charges [55]			Lennard-Jones potential		
Symbol	**Weight [u]**	**Charges [e]**	**Bond**	D_{ij} **(eV)**	R_{ij} **(nm)**
Oh	16.00	−1.40	O—O	0.15540	0.35532
Ho	1.00	0.40	O—Ca	3.83368×10^{-5}	0.48980
Ca	39.90	2.00	Ca—Ca	2.18105×10^{-7}	0.62428
Oh_w	16.00	−0.95	O—H	6.73854×10^{-3}	0.35532
Ho_w	1.00	0.42	H—H	0.02000	0.20000

Table 5.
CH parameters of atomic charges and atomic force field employed (1e = 1.6021892 × 10^{-19} C).

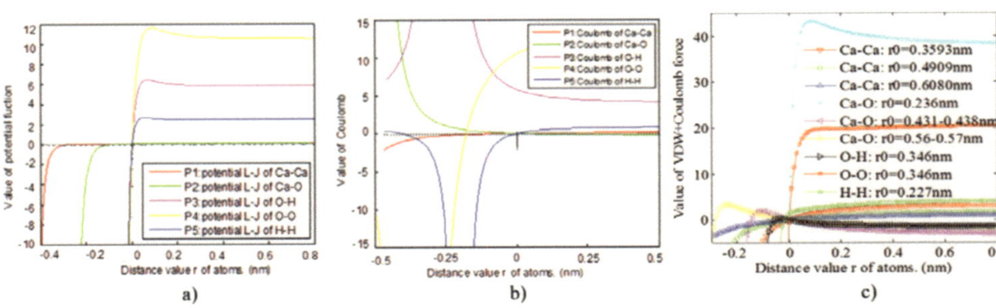

Figure 16.
Ionic bond's definition and various atomic types of CH by AFEM modeling. (a) VDW force interaction. (b) Coulomb force of interaction. (c) Coupling force field.

Figure 17 shows the stretching schematic diagram of CH crystal in three directions. The CH crystal modeling based on AFEM is extensively concerned in this section. CH crystal modeling using the equivalent beam-spring element is mainly investigated, and elastic modulus of CH hexagonal polycrystals can be finally determined based on the calculated elastic constants matrix.

More particularly, the results of the AFEM are compared to previous results of experiment [56]. Comparison of CH stress-strain and elastic modulus by AFEM modeling is shown in **Figure 18**.

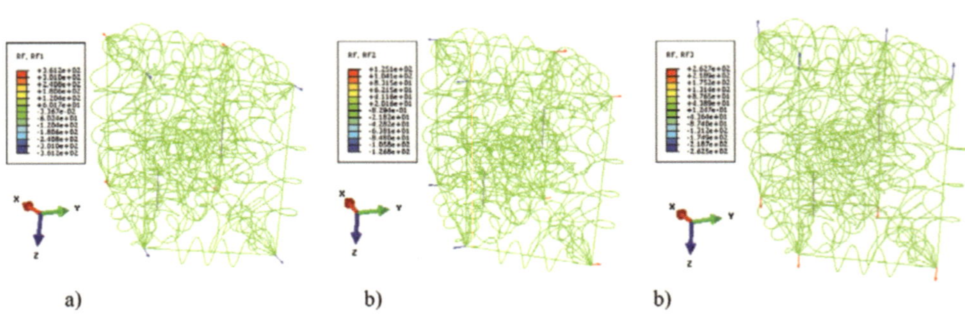

Figure 17.
Strain loading of CH crystal. (a) Strain loading in x-direction. (b) Strain loading in y-direction. (c) Strain loading in z-direction.

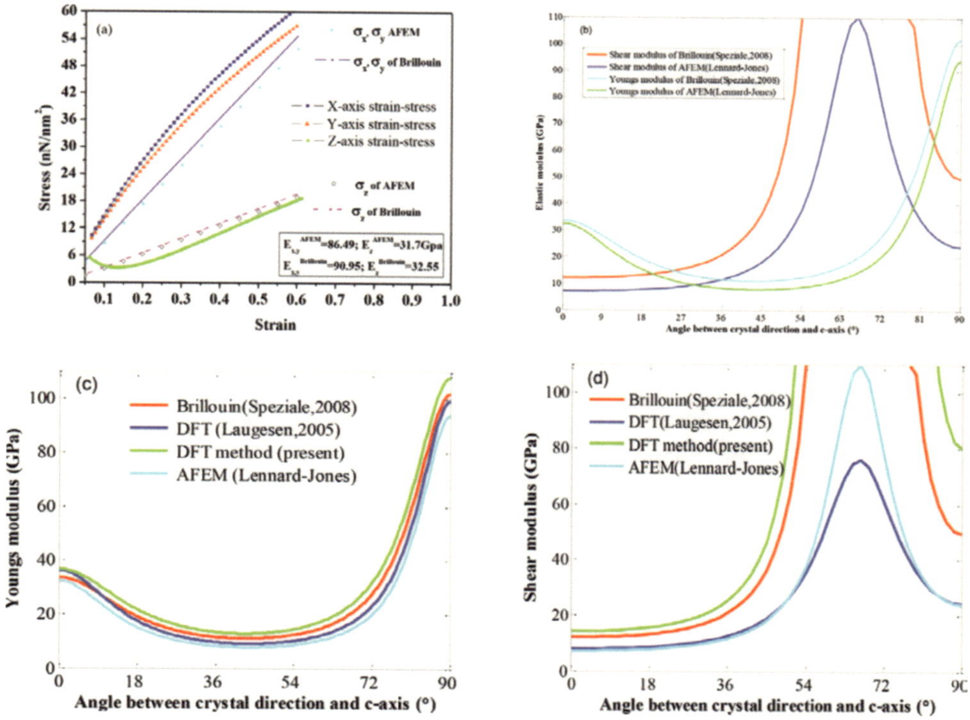

Figure 18.
Stress-strain and comparison of elastic moduli by various methods. (a) Stress-strain and axial modulus by AFEM. (b) Comparison of Young's and shear moduli. (c) Comparison of Young's modulus of single CH crystal. (d) Comparison of shear modulus of single CH crystal.

From **Figure 18(a)**, it can be seen that, even if all the results are in quite good agreement, the numerical results using AFEM are slightly lower than those obtained by Brillouin spectroscopy measurement. Anisotropy of elastic modulus in each direction with various angles of z-axis is shown in **Figure 18(b)**. The comparison of elastic modulus curves based on C_{ij} shows anisotropy in each direction with various angles of z-axis in **Figure 18(c)**. As can be seen from **Figure 18(d)** and axial modulus formulas, even if all the results are in quite good agreement, it can be noted that DFT results are upper than those of Brillouin spectroscopy measurement, whereas AFEM ones are lower.

Elastic constant C_{ij} is then calculated and axial modulus can be obtained by Eqs. (45) and (47). Axial moduli based on elastic constants by AFEM are as: $\sigma_x = \sigma_y = 86.4918\varepsilon_x$, $\sigma_z = 31.7002\varepsilon_z$. From **Figure 18**, shear and Young's moduli by AFEM are slightly lower than Brillouin spectroscopy measurement. Moreover, elastic moduli by Wu et al. [38] are as: $B = 28.249$ GPa, $G = 17.314$ GPa. Thus, Young's modulus of CH polycrystals is averaged to be 43.130 GPa by Reuss-Voigt-Hill estimation, which is in quite good agreement with experimental averaged value (39.88 GPa) and with the literature values (45.94 GPa by Laugesen, 52.4 GPa by Speziale, 44.69 GPa by Kerisit, and 46.58 GPa by Holuj). AFEM has a certain meaning to provide a method to calculate elastic modulus of other hexagonal structures.

Strain-stress data and elastic moduli (bulk, shear, and Young's moduli) by AFEM are given in **Tables 6** and 7. Strain-stress data of CH crystal by AFEM are given in **Table 6**. Elastic modulus of polycrystalline CH compared with previous results of the literature is given in **Table 7**.

	Strain-stress by volume unit anisotropy				Strain-stress by beam-spring element		
σ_{ij} (nN/nm^2)	ε_{11}	ε_{22}	ε_{33}	σ_{ij} (nN/nm^2)	ε_{11}	ε_{22}	ε_{33}
σ_{11} = 74.3832	0.834493	−0.249651	−0.118612	σ_{11} = 41.0121	0.453152	−0.064053	−0.052899
σ_{22} = 59.7722	−0.200612	0.670576	−0.095314	σ_{22} = 32.3671	−0.057148	0.337453	−0.052226
σ_{33} = 36.1151	−0.057590	−0.057590	1.018542	σ_{33} = 11.3929	−0.041024	−0.041024	0.3483412
σ_{12} = 4.0644	γ_{12} = 0.515788			σ_{12} = 3.5601	γ_{12} = 0.501512		

Table 6.
Strain-stress data of CH by volume unit using FEM and beam-spring element using AFEM modeling.

DOI: http://dx.doi.org/10.5772/intechopen.84597

Method	C_{11}	C_{12}	C_{13}	C_{33}	C_{44}	C_{66}	*Bv*	*Br*	*Gv*	*Gr*	μ
FEM	99.38	30.78	7.36	36.29	7.88	34.31	36.23	26.63	22.65	13.92	0.2
AFEM	94.02	25.51	6.02	32.31	7.099	34.25	32.83	23.67	21.88	12.75	0.246
Equation-AFEM	95.806	25.294	4.626	35.671	7.098	35.256	32.93	24.59	22.74	12.99	0.243
Equation-FEM	100.22	31.611	6.389	37.430	7.878	34.305	36.29	26.79	22.91	14.00	0.255
Brillouin [56]	102.00	32.00	8.40	33.60	12.00	35.00	37.24	26.02	24.39	18.18	0.225

Table 7.
Elastic constants of CH and homogenized moduli (bulk, shear, and Young's moduli) of polycrystalline CH by FEM and AFEM (unit: GPa).

6. Conclusions

Equivalent beam elements are applied in zigzag nanotubes of C—C covalent bond, as well as the AFEM modeling of a tensile test on CH and TLGSs. These models at nanoscale are established and then elastic constants are determined based on several methods/algorithm existed and developed. Conclusions are as:

1) A theoretical formula between axial modulus $E_{x,y,z}$ and c_{ij} of hexagonal crystal is derived by crystal elastic theory. Besides, a reliable equation to calculate these C_{ij} is also given out and verified by both volume unit (FEM) and equivalent beam-nonlinear spring elements (AFEM), which are as follows:

   a) The expressions of the x-axial modulus E_1, y-axial modulus E_2, and z-axial modulus E_3 of hexagonal crystal as a function of elastic constants C_{ij} are as:

$$E_3 = c_{33} - \frac{2c_{13}^2}{c_{11} + c_{12}}; \quad E_1 = E_2 = (c_{11} - c_{12})\left[1 + \left(c_{12}c_{33} - c_{13}^2\right)/\left(c_{11}c_{33} - c_{13}^2\right)\right]$$

   b) The equation of elastic constants Cij calculated by the stress components and strain components using the volume unit (FEM method) are derived and detailed as in Eq. (43), where $A = \varepsilon_{11}$, $B = \varepsilon_{22}$, $C = \varepsilon_{33}$, $D = \varepsilon_{31}$, $E = \varepsilon_{11} + \varepsilon_{21}$ (ε_{ij} are independent strains in all directions).

2) The product $E.t$ of Young's modulus and wall thickness for CNT is mainly distributed in the constant range of 0.33–0.37 kJ/m^2. Young's modulus of zigzag SWCNT is 1.05 Tpa and $E.t$ is close to be 0.3604 kJ/m^2. Besides, the calculated E of TLGSs model is close to the measurement of Berryman.

3) Young's modulus of single-layer graphene calculated by AFEM is 1025.81 GPa, compared with 996.49 GPa by Berryman, and the relative errors of c_{ij} are in an acceptable range with experimental values.

4) CH crystal has shown anisotropy. Based on the stress-strain data of AFEM modeling, elastic constant c_{ij} is calculated, Young's modulus is estimated to be 43.130 GPa using Reuss-Voigt-Hill estimation, which is close to the Brillouin spectroscopy result of 52.15 GPa.

As an attempt, the computational efficiency of AFEM method of CH crystal on ABAQUS software is greatly improved. Besides, the computation cycle (no more than 5 min in total) is greatly reduced. Thus, based on the content above, AFEM can be considered as a kind of alternative method to obtain mechanical properties of crystals at nanoscale.

However, it should be noted that molecular mechanics methods cannot provide information of the electronic structure as they omit the electronic nature of the atoms. That is why the first-principles simulation (the density functional theory, etc.) is also introduced to calculate the Young's moduli of other typical structures considered in this thesis, under the situation of lacking a complete potential function parameters or more complex chemical bonds existed in the structures.

In short, we can see that AFEM modeling methodology at the atomic scale (or nanoscale) can also be an effective method for the specific models, since the macroscopic properties depend largely on the physicochemical properties of the interatomic bonds. Multiscale methods, including the AFEM method, therefore connect nanoscale modeling and continuous pattern of deformation behavior by identifying relevant parameters from small scales to larger scales, thus have their prospective applications. It is not far for us to deeply investigate mechanical properties with the rapid development of computer hardware level, the interatomic potential functions accurately described, new high-efficiency algorithms, etc. Perspectives of the computer tools, accurate interatomic potential functions, and new high-efficiency algorithms will complete this meaningful work and study the mechanical behavior of materials more deeply.

Acknowledgements

The author acknowledges the financial support provided by the China Scholarship Council and the startup foundation of Xi'an Shiyou University. Thanks Qiufeng Wang for her proofreading, and very sincere thankness to prof. Fabrice Bernard and prof. Siham Kamali-Bernard for the their enthusiastic and responsible guideness.

Author details

Jia Fu[1,2]

1 Xi'an Shiyou University, Xi'an, China

2 INSA de Rennes, Rennes, France

*Address all correspondence to: fujia@xsyu.edu.cn

References

[1] Sarma SD, Geim AK. Foreword. Reprint of: Foreword. Solid State Communications. 2007;**143**(27–28):1-2

[2] Jiao LY, Zhang L, Wang XR, Diankov G, Dai HJ. Narrow graphene nanoribbons from carbon nanotubes. Nature. 2009;**458**:877-880

[3] Treacy MMJ, Ebbesen TW, Gibson JM. Exceptionally high young's modulus observed for individual carbon nanotubes. Nature. 1996;**381**(6584): 678-680

[4] Shokrieh MM, Rafiee R. Prediction of Young's modulus of graphene sheets and carbon nanotubes using nanoscale continuum mechanics approach. Materials and Design. 2010;**31**:790-795

[5] Yakobson BI, Brabec CJ, Bernholc J. Nanomechanics of carbon tubes: Instabilities beyond linear response. Physical Review Letters. 1996;**76**(14): 2511

[6] Vodenitcharova T, Zhang LC. Effective wall thickness of a single-walled carbon nanotube. Physical Review B. 2003;**68**(16):165401

[7] Zhou X, Zhou JJ, Ou-Yang ZC. Strain energy and Young's modulus of single-wall carbon nanotubes calculated from electronic energy-band theory. Physical Review B. 2000;**62**(20):13692

[8] Li C, Chou T-W. A structural mechanics approach for the analysis of carbon nanotubes. International Journal of Solids and Structures. 2003;**40**: 2487-2499

[9] Goze C, Vaccarini L, Henrard L, et al. Elastic and mechanical properties of carbon nanotubes. Synthetic Metals. 1999;**103**(1):2500-2501

[10] Popov VN, Van Doren VE, Balkanski M. Elastic properties of single-walled carbon nanotubes. Physical Review B. 2000;**61**:3078-3084

[11] Kudin KN, Scuseria GE. C2F, BN, and C nanoshell elasticity from ab initio computations. Physical Review B. 2001; **64**(23):235-406

[12] Tu Z, Ou-Yang Z. Single-walled and multiwalled carbon nanotubes viewed as elastic tubes with the effective Young's moduli dependent on layer number. Physical Review B. 2002;**65**: 233-407

[13] Chang T, Gao H. Size-dependent elastic properties of a single-walled carbon nanotube via a molecular mechanics model. Journal of the Mechanics and Physics of Solids. 2003; **51**:1059-1074

[14] Sears A, Batra RC. Macroscopic properties of carbon nanotubes frommolecular-mechanics simulations. Physical Review B. 2004;**69**(23):235406

[15] Pantano A, Boyce MC, Parks DM. Mechanics of axial compression of single and multi-wall carbon nanotubes. Journal of Engineering Materials and Technology. 2004;**126**(3):279-284

[16] Zhang HW, Wang JB, Guo X. Predicting the elastic properties of single-walled carbon nanotubes. Journal of the Mechanics and Physics of Solids. 2005;**53**(9):1929-1950

[17] To CWS. Bending and shear moduli of single-walled carbon nanotubes. Finite Elements in Analysis and Design. 2006;**42**:404-413

[18] Chandraseker K, Mukherjee S. Atomistic-continuum and ab initio estimation of the elastic moduli of single-walled carbon nanotubes. Computational Materials Science. 2007; **40**:147-158

[19] Wang CY, Zhang LC. An elastic shell model for characterizing single-walled carbon nanotubes. Nanotechnology. 2008;**19**:195704

[20] Lu J, Chen H. Analysis of single-walled carbon nanotubes using a chemical bond element model. Chinese Journal of Chemical Physics. 2008; **21**(4):353-360

[21] Lu X, Zhong H. Mechanical property evaluation of single-walled carbon nanotubes by finite element modeling. Composites: Part B. 2012;**43**: 1902-1913

[22] Neugebauer R, Wertheim R, Semmler U. The atomic finite element method as a bridge between molecular dynamics and continuum mechanics. Journal of Multiscale Modelling. 2011;**3** (01–02):39-47

[23] Rappe AK, Casewit CJ, et al. UFF: A full periodic-table force-field for molecular mechanics and molecular dynamics simulations. Journal of the American Chemical Society. 1992;**114**: 10024-10035

[24] Kwon YK, Berber S, Tománek D. Thermal contraction of carbon fullerenes and nanotubes. Physical Review Letters. 2004;**92**(1):015901

[25] Leach AR. Molecular Modelling: Principles and Applications. 2nd ed. Harlow: Pearson education, England; 2001

[26] Cramer CJ. Essentials of Computationa Chemistry: Theories and Models. Chichester, England: John Wiley & Sons; 2002

[27] Jensen F. Introduction to Computational Chemistry. Vol. xvi. Chichester; New York: Wiley; 1999. 429p

[28] Jia F, Bernard F, Kamali-Bernard S. Nanoscale modeling and elastic properties of Portlandite and graphene based on atomic finite element method. Applied Mechanics and Materials. 2015; **711**:137-142

[29] Hirschfelder JO, Curtiss CF, Bird RB. Molecular Theory of Gases and Liquids (2nd Corrected Printing). New York: John Wiley and Sons; 1964

[30] Li CY, Chou TW. Elastic moduli of multi-walled carbon nanotubes and effect of van der Waals forces. Composites Science and Technology. 2003;**63**:1517-1524

[31] Frenkel D, Smit B. Understanding Molecular Simulation: From Algorithm to Application. 2nd ed. San Diego: Academic Press; 2002

[32] Reed TM, Gubbins KE. Applied Statistical Mechanics. New York: McGraw-Hill; 1973

[33] Liu B, Jiang H, Huang Y, Qu S, Yu MF, et al. Atomic-scale finite element method in multiscale computation with applications to carbon nanotubes. Physical Review B. 2005; **72**(3):035435

[34] Kurukuri S, Eckardt DIS. A review of homogenization techniques for heterogeneous materials. Term Paper. Advanced Mechanics of Materials and Structures. Germany: Graduate School in Structural Engineering; 2004

[35] Yan CK. On homogenization and de-homogenization of composite materials. Drexel University; 2003

[36] Abaqus Users Manual. Version 6.13-2, Internet Manual of Simulia. Rhode Island, USA: Dassault Systémes Simulia Corp., Providence; 2013

[37] Garboczi EJ. Finite element and finite difference programs for computing the linear electric and elastic properties of digital images of random

materials. NIST Internal Report 6269; 1998

[38] Wu ZJ, Zhao EJ, Xiang HP, et al. Crystal structures and elastic properties of superhard IrN2 and IrN3 from first principles. Physical Review B. 2007; **76**(5):054115

[39] Girifalco LA, Hodak M, Lee RS. Carbon nanotubes, buckyballs, ropes, and a universal graphitic potential. Physical Review B. 2000;**62**(19): L104-L131

[40] Jia F, Bernard F, Kamali-Bernard S. Multiscale modeling and mechanical properties of zigzag CNT and triple-layer graphene sheet based on atomic finite element method. Journal of Nano Research. 2015;**33**:92-105

[41] Brenner DW. The art and science of an analytic potential. Physica Status Solidi B. 2000;**217**(1):23-40

[42] Brenner DW, Shenderova OA, Harrison JA, et al. A second-generation reactive empirical bond order (REBO) potential energy expression for hydrocarbons. Journal of Physics: Condensed Matter. 2002;**14**(4):783-802

[43] Chen X, Cao G. A structural mechanics study of single-walled carbon nanotubes generalized from atomistic simulation. Nanotechnology. 2006; **17**(4):1004

[44] Lu WB, Wu J, Song J, et al. A cohesive law for interfaces between multi-wall carbon nanotubes and polymers due to the van der Waals interactions. Computer Methods in Applied Mechanics and Engineering. 2008;**197**(41):3261-3267

[45] Berryman JG. Bounds and self-consistent estimates for elastic constants of random polycrystals with hexagonal, trigonal, and tetragonal symmetries. Journal of the Mechanics and Physics of Solids. 2005;**53**(10):2141-2173

[46] Berryman JG. Poroelastic shear modulus dependence on pore-fluid properties arising in a model of thin isotropic layers. Geophysical Journal International. 2004;**157**(1):415-425

[47] Berryman JG. Bounds on elastic constants for random polycrystals of laminates. Journal of Applied Physics. 2004;**96**(8):4281-4287

[48] Berryman JG. Bounds and self-consistent estimates for elastic constants of polycrystals composed of orthorhombics or crystals with higher symmetries. Physical Review E. 2011; **83**(4):046130

[49] Neek-Amal M, Peeters FM. Nanoindentation of a circular sheet of bilayer graphene. Physical Review B. 2010;**81**(23):235421

[50] Kamali-Bernard S, Bernard F. Effect of tensile cracking on diffusivity of mortar: 3D numerical modelling. Computational Materials Science. 2009; **47**:178-185

[51] Bernard F, Kamali-Bernard S, Prince W. 3D multi-scale modelling of mechanical behavior of sound and leached mortar. Cement and Concrete Research. 2008;**38**:449-458

[52] Jia F, Bernard F, Kamali-Bernard S. First-principles calculations of typical anisotropic cubic and hexagonal structures and homogenized moduli estimation based on the Y-parameter. Application to CaO, MgO, CH and calcite $CaCO_3$. Journal of Physics and Chemistry of Solids. 2017;**101**:74-89

[53] Laugesen JL. Density functional calculations of elastic properties of portlandite, $Ca(OH)_2$. Cement and Concrete Research. 2005;**35**(2):199-202

[54] Mitev PD, Bopp PA, Petreska J, et al. Different structures give similar vibrational spectra: The case of OH^- in

aqueous solution. The Journal of Chemical Physics. 2013;**138**(6):064503

[55] Galmarini S, Aimable A, Ruffray N, Bowen P. Changes in portlandite morphology with solvent composition: Atomistic simulations and experiment. Cement and Concrete Research. 2011; **41**(12):1330-1338

[56] Speziale S, Reichmann HJ, Schilling FR, Wenk HR, et al. Determination of the elastic constants of portlandite by Brillouin spectroscopy. Cement and Concrete Research. 2008;**38**(10): 1148-1153

Chapter 3

Classical Density Functional Theory (DFT): Elastic Constants of Typical Single Crystals

Jia Fu

Abstract

Anisotropy becomes the focus of condensed matter physics, materials physics, and materials chemistry research, where molecular pressure characteristics are critical to better understand the mechanical properties of the materials at the atomic scale. So the influencing effect of pressure on structural, elastic properties of single cubic crystals (CaO and MgO), single hexagonal crystals (CH and calcite), and single monoclinic crystals (11 Å tobermorite and gypsum) are mainly investigated. This chapter is committed to model and calculate elastic constants of typical anisotropic crystal structures. Above all, the functional local density approximation (LDA) or generalized gradient approximation (GGA) is mainly used. Then, the elastic constants are calculated, which can be used as the measure criterion of the resistance of a crystal to an externally applied stress. Besides, the optimized structural lattice parameters at zero pressure are calculated to compare with the experimental parameters, which all have good agreement with the experimental and theoretical values. Therefore, it is fairly meaningful to study the elastic constants to understand the physical, chemical, and mechanical properties of these cubic, hexagonal, and monoclinic crystal structures. The results show that the applied pressure is beneficial to the elastic properties. These researches provide basic physical parameters for their homogenized moduli of these typical polycrystals.

Keywords: DFT, lattice parameters, elastic constants, GGA, LDA, pressure effect

1. Introduction

Anisotropy can effectively increase the freedom of material, thus can be used to provide controllability and design new materials, and anisotropy becomes the focus of condensed matter physics, materials physics, and materials chemistry research. Molecular pressure characteristics are critical to better understand the mechanical properties of the materials at the atomic scale.

First-principles simulation starting from quantum mechanics is a powerful tool for materials research, of which the density functional theory (DFT) is commonly used to study the crystal structure, lattice energy, the equation of state, the electronic bandgap, and vibration spectral properties [1]. DFT is confirmed to be one of the important nanoscale modeling methods. With the increasing interest in the advanced applications of cementitious materials at the nanoscale, elastic constants of single crystal are essential to the material performance because many other parameters related to mechanical properties can be derived from them. This resistant force of bond in crystal can be characterized by energy gap, and the number of bond per unit area can be determined by valence electron density.

DFT has a great importance of simplifying the solving process of the complex Schrödinger equation. The internal degrees of freedom of the nuclei extend to a scale of several of magnitude orders (smaller than the electron variables and concentrated at the bulk of the mass), which causes a significant slowdown compared to electrons. Based on the kinetic energy density functional of Thomas [2] and the exchange-correlation effects of Dirac [3], the density functional theory has been greatly developed by Kohn and Sham [4], who have established the fundamental approximation theorem on the functional status to describe real systems by electronic structure calculations. Assuming that the validity of the exchange-correlation functional and the accuracy of density and energy convergence are correct, we should also clear that the eigenvalues of KS equations have no physical meaning and the ionization energy is in the opposite state direction [5].

Many studies show that the band structure obtained by KS equations remains more reasonable prediction of experimental energy spectra in the main usage of functional local density approximation (LDA) or generalized gradient approximation (GGA). Moreover, the use of the potential and the disturbance of eigenfunctions allow us to determine the excitation energies during KS solving process. One proposed approach is to introduce the eigenstates to calculate multibody (many-body calculation) on the basis of Monte Carlo calculations [6] and perturbation theory [7]. The approach of Kohn and Sham (KS) [4] proposed in 1965 is still the most commonly used approach in DFT calculations, which aims to determine the specific properties of a multiparticle system using independent particle and owns satisfactory success.

So, we mainly investigate some typical crystals (such as: cubic CaO and MgO, hexagonal portlandite and calcite, monoclinic 11 Å tobermorite and gypsum), focusing on the elastic constants within a certain pressure range, so as to lay a foundation for the elastic modulus calculations.

2. Molecular modeling using DFT by materials studio

2.1 Introduction of density functional theory (DFT)

From a microscopic point of view, Schrödinger equation describing a periodic crystal system composed of atomic nuclei n in mutual interaction and electron spin σ_i is positioned $\vec{R}=\left\{\vec{R_i}; i=1, \ldots N_n\right\}$ and $\vec{r}=\left\{\left(\vec{r_i}, \sigma_i\right); i=1, \ldots N_e\right\}$, respectively.

$$H\psi\left(\vec{R}, \vec{r}\right) = E\psi\left(\vec{R}, \vec{r}\right) \tag{1}$$

The possible analytical representation and resolution of such a problem becomes a difficult task due to the limited memory of the computer tools. But it is possible to reformulate the problem using appropriate theorems and approximations.

A multiparticle electronic structure satisfies the Schrödinger equation. Generally, Hamiltonian contains kinetic energy, Coulomb potential energy between electrons and the external potential. Kohn-Sham equation as an approximation to simplify Schrödinger equation is described. For crystal composed by vibrator with the vibration frequency w_i, the total Helmholtz free energy is:

$$F = E + \sum_i F_i^{th} = U + \sum_i \frac{1}{2}\hbar w_i + k_B T \sum_i In\left(1 - e^{-\frac{\hbar w_i}{k_B T}}\right) \tag{2}$$

Helmholtz free energy can be calculated for all the thermodynamic quantities. DFT-QHA (quasi-harmonic approximation) is a precise calculation method to calculate thermodynamic properties of solid material elastic constants and Debye temperature with the accurate predictions. Therefore, DFT-QHA method is quite suitable to study physical properties of crystal structure under high pressure.

The Hamiltonian and Kohn-Sham equations as an approximation to simplification of Schrodinger equation to solve the multiparticle systems are [8]:

$$\hat{H} = -\frac{\hbar}{2m_e}\sum_i \nabla_i^2 + \sum_i V_{ex}(r_i) + \frac{1}{2}\sum_{i \neq j}\frac{e^2}{|r_i - r_j|} \tag{3}$$

$$\begin{cases} \left[-\frac{\hbar}{2m}\nabla^2 + V_{eff}\left(\vec{r}\right)\right]\varphi_i\left(\vec{r}\right) = \varepsilon_i \varphi_i\left(\vec{r}\right) \\ V_{eff}(r) = V\left(\vec{r}\right) + V_{ee}\left(\vec{r}\right) + V_{ex}\left(\vec{r}\right) \\ V_{xe}\left(\vec{r}\right) = \frac{\delta E_{xe}[\rho]}{\delta\left(\vec{r}\right)} \\ \rho\left(\vec{r}\right) = \sum_i \left|\phi_i\left(\vec{r}\right)\right|^2 \end{cases} \tag{4}$$

where $V_{ex}(r)$ is an external potential; r_i and r_j are the nucleus position vector; while m_ε, m, respectively, stand for the quality of nuclei and electrons; $\rho(r)$ is the electron density. Eqs. (3) and (4) include not only the nucleus and electron kinetic energy term, but also interaction term of nucleus-nucleus and electron-electron [8].

According to the theory of elasticity, under the isothermal strain, the elastic modulus of Helmholtz free energy can be described by the form of the Taylor expansion, of which the coefficients of the polynomial is the elastic coefficient:

$$\rho_0 F\left(\eta_{ij}, T\right) = \rho_0 F\left(\eta_{ij}, T\right) + \frac{1}{2}\sum_{ijkl} c_{ijkl}^T \eta_{ij}\eta_{kl} + \dots + \frac{1}{n!}\sum_{ijkl} c_{ijkl\dots}^T \eta_{ij}\eta_{kl}\dots \tag{5}$$

where η_{ij}, η_{kl}, and η_{mn} are the coefficients of Lagrange deformation tensor, c_{ijkl}^T is the isothermal first-order elastic coefficients, and $F\left(\eta_{ij}, T\right)$ is the Helmholtz free energy.

The components of the stress tensor can be extracted by $\sigma_i = \sum_{j=1}^{6} c_{ij}\varepsilon_j$, after the applied strain, the total energy variation of the system can be expressed as:

$$\Delta E = \frac{V}{2}\sum_{i=1}^{6}\sum_{j=1}^{6} c_{ij} e_i e_j \tag{6}$$

As the independent elastic constants of different crystals are different from each other, through the different matrix elements of the crystal elastic modulus, we can get different strain applied methods.

2.2 Introduction of the DFT and CASTEP module

At present, there are a lot of software for molecular modeling, such as: Ceruis2, InsightII, VASP, Gaussian, Material Studio, etc., of which Ceruis2, InsightII run under Unix mainframe systems and thus are inconvenient to use. Material Studio can run a set of quantum mechanics, molecular mechanics, mesoscopic model, and statistical correlation simulation. Moreover, Materials Studio has an environment of server/client mode, which has brought the advanced material simulation and modeling techniques.

Materials Studio CASTEP has been applied to a wide range of research problems such as surface chemistry, and chemisorption, heterogeneous catalysis, defects in semiconductors, grain boundaries, stacking faults, nanotechnology, molecular crystals, polymorphic studies, diffusion mechanisms, and molecular dynamics of liquids. Materials Studio CASTEP uses a total energy plane-wave pseudopotential method. In the mathematical model of the material, Materials Studio CASTEP replaces core electrons with effective potentials acting only on the valence electrons in the system. Electronic wavefunction are expanded through a plane-wave basis set, and exchange and correlation effects can be included within either the local density approximation (LDA) or generalized gradient approximation (GGA). Combining the use of pseudopotentials and plane wave basis sets enables extremely efficient geometry optimizations of molecules, solids, surfaces, and interfaces. Based on total energy pseudopotential methods, Materials Studio CASTEP can predict properties such as lattice constants, molecular geometry, elastic constants, band structures, density-of-states, charge densities and wave functions, and optical properties. The pseudopotential plane-wave technology underlying CASTEP is well validated, and efficient parallel versions of the code are also available for large systems involving hundreds of atoms.

CASTEP is capable of computing many electronic and optical properties using density functional perturbation theory (DFPT), known as the linear response method. The primary reason that CASTEP has become so powerful is that the numerical methods used to solve the underlying quantum mechanical calculations are both computationally efficient and extremely accurate. This approach makes possible a wider variety of properties than are possible using the so-called finite difference approaches, which require repeated computations on series systems.

For a detailed description of the quality settings, we can set the calculation parameters as in **Table 1**.

Convergence criteria of the accuracy of the calculation are controlled in **Table 1** by the quality settings. Specify the number of steps for each strain and maximum strain amplitude in the appropriate text boxes. The minimizer tab of the CASTEP geometry optimization dialog is used to obtain the optimized structure. Specify the quality using either the dropdown list or by providing individual convergence tolerances for the energy, maximum force, and maximum displacement in the corresponding text boxes.

Value	Coarse	Medium	Fine	Ultrafine
SCF tolerance (eV/atom)	1.0×10^{-5}	2.0×10^{-6}	1.0×10^{-6}	5.0×10^{-7}
k-Points separation (1/Å)	0.07	0.05	0.04	0.04
Energy cutoff	Coarse	Medium	340 eV	380 eV
Energy tolerance (eV/atom)	5.0×10^{-5}	2.0×10^{-5}	1.0×10^{-5}	5.0×10^{-6}
Max. force tolerance (eV/Å)	0.1	0.05	0.03	0.01
Max. stress tolerance (GPa)	0.2	0.1	0.05	0.02
Max. displacement tolerance (Å)	5.0×10^{-3}	2.0×10^{-3}	1.0×10^{-3}	5.0×10^{-4}

Table 1.
Parameter accuracy of CASTEP module.

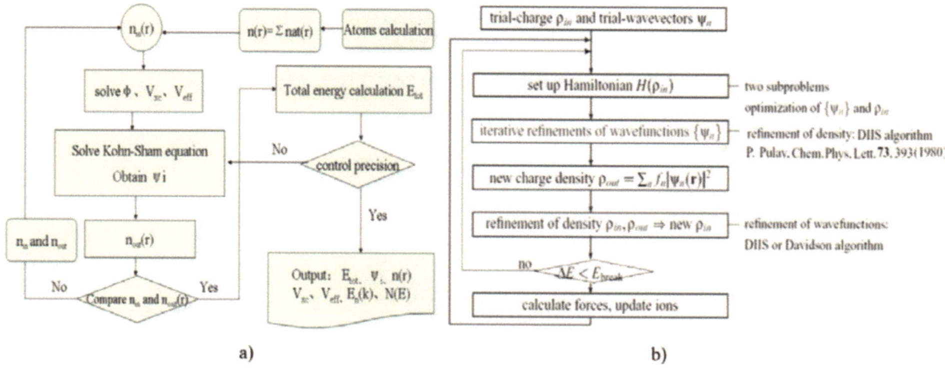

Figure 1.
Solution of Kohn-Sham equations. (a) Theoretical calculation of Kohn-Sham equations. (b) Flowchart of solving Kohn-Sham equations by DFT.

Elastic constants will be generated for the theoretical lattice constants after cell optimization. Theoretical calculation of Kohn-Sham equations is shown in **Figure 1(a)**, and the corresponding DFT calculation process is shown in **Figure 1(b)**.

From **Figure 1**, the calculation of elastic constants is proceeded by full geometry optimization, including cell optimization. Elastic constants are evaluated by calculating the stress tensor for a number of distorted structures. Internal coordinates are optimized in each run, while keeping the lattice parameters fixed. The accuracy of the elastic constants depends to a great extent on the accuracy of the self-consistent field (SCF) part and also on the level of convergence of geometry optimizations for each distorted structure.

3. Theoretical approximate solution of DFT calculation

The crystalline ion movement of the electron is as: $\psi\left(\vec{R},\vec{r}\right)=\chi\left(\vec{R}\right)\varphi\left(\vec{R},\vec{r}\right)$ and assume that the electron mobility (φ) does not depend on the speed nuclei but on their positions.

3.1 Solving equation of adiabatic approximation

According to the Born-Oppenheimer or adiabatic approximation [9], the dynamics of the system (electrons and nuclei) is described. The electrons are assumed to react instantly to ionic motion. Therefore, in electronic coordinates, the positions of the nucleus are considered immobile external parameters.

$$\hat{H}=\hat{T}_e+\hat{U}_{ne}+\hat{U}_{ee}+\hat{U}_{nn} \tag{7}$$

$$\hat{H}_{\vec{R}}\varphi^0_{\vec{R}}\left(\vec{r}\right)=E^{BO}\left(\vec{R}\right)\varphi^0_{\vec{R}}\left(\vec{r}\right)\left(\vec{r}\right) \tag{8}$$

where the last term of the Hamiltonian is constant and has been introduced in order to preserve the neutrality of the system and avoid the divergence of the eigenvalues. Clean the ground state of the system for fixed nuclear positions, total energy is given by the formula:

$$E^{BO}\left(\vec{R}\right)\left\langle\varphi^0_{\vec{R}}|\hat{H}|\varphi^0_{\vec{R}}\right\rangle=\min\left\langle\varphi^0_{\vec{R}}|\hat{H}|\varphi^0_{\vec{R}}\right\rangle \tag{9}$$

This energy has a surface in the space coordinates that is said ionic Born Oppenheimer surface.

The ions move according to the effective potential energy, including Coulomb repulsion and the anchoring effect of the electron, as:

$$\hat{H}^{BO} = \hat{T}_n + E^{BO}\left(\vec{R}\right) \tag{10}$$

$$\hat{H}^{BO}\chi\left(\vec{R}\right) = E\chi\left(\vec{R}\right) \tag{11}$$

The dissociation degrees of freedom of electrons from those of nucleons, obtained through the adiabatic approximation, are very important, because if the electrons must be treated by quantum mechanics, degrees of freedom of ions in most cases are processed in a conventional manner.

This formulation applies to any system of mutually interacting particles in an external potential $V_{ex}\left(\vec{r}\right)$, where the Hamiltonian is written as:

$$\hat{H} = -\frac{\hbar}{2m_e}\sum_i \nabla_i^2 + \sum_i V_{ex}(r_i) + \frac{1}{2}\sum_{i \neq j}\frac{e^2}{|r_i - r_j|} \tag{12}$$

DFT and its founding principle is summarized in two theorems, first introduced by Hohenberg and Kohn, which say there is bijection between the set of potential $V_{ex}\left(\vec{r_i}\right)$ and the density minimizing the Eq. (9), based on the following points:

The total energy of the ground state of a system for interacting electrons is functional (unknown) of the single electron density.

$$E_{HK}[n] = T[n] + E_{\text{int}}[n] + \int d^3r V_{ex}(r) + E_{nm}\left(\vec{R}\right) \cdot F_{HK}[n] + \int d^3r V_{ex}(r) + E_{nn}\left(\vec{R}\right) \tag{13}$$

As a result, the density $n_0(r)$ minimizing the energy associated with the Hamiltonian in Eqs. (13) is obtained and used to evaluate the energy of the ground state of the system.

The principle established in the second theorem of Hohenberg and Kohn specifies that the density that minimizes the energy is the energy of the ground state.

$$E^{BO}\left[\vec{R}\right] = \min E\left(\vec{R}, n\left(\vec{r}\right)\right) \tag{14}$$

Because the ground state is concerned, it is possible to replace (in the Hilbert space) the wave system function by the electron charge density, which therefore becomes the fundamental quantity of the problem. In principle, the problem boils down to minimize the total energy of the system in accordance with the variations in the density governed by the constraint on the number of particles $\int n\left(\vec{r}\right) d^3r = N_e$. In this stage, the DFT can reformulate the problem rather than solve an uncertain functional $F_{HK}(n)$.

3.2 The approximation approach of Kohn-Sham

This approach assumes that the density in the ground state of the system is equal to that in some systems composed of noninteracting particles. This involves independent particle equations for the noninteracting (numerically manageable) system, gathering all the terms complicated and difficult to assess, in a functional exchange-correlation $E_{XC}(n)$.

$$E_{KS} = F[n] + \int d^3 r V_{ex}(r) = T_S[n] + E_H[n] + E_{XC}[n] + \int d^3 r V_{ex}(r) \tag{15}$$

T is the kinetic energy of a system of particles (electrons) independent (noninteracting) embedded in an effective potential which is no other than the real system,

$$T_S[n] = \langle \psi_{NI} | \hat{T}_e | \psi_{NI} \rangle = \sum_{i=1}^{N_e} \left\langle \varphi_i \middle| -\frac{1}{2}\nabla^2 \middle| \varphi_i \right\rangle \tag{16}$$

The Hartree energy or energy of interaction is associated with the Coulomb interaction of the self-defined electron density

$$E_{Hartree}[n] = \frac{1}{2}\int d^3 r d^3 r' \frac{n(r)n(r')}{|r - r'|} \tag{17}$$

$$n(r) = \sum_{i=1}^{N_e} |\varphi_i(r)|^2 \tag{18}$$

Solving the auxiliary Kohn and Sham system for the ground state can be seen as a minimization problem while respecting the density $n(r)$. Apart from that T_S is orbital function, all other terms depend on the density. Therefore, it is possible to vary the functions of the wave and deriving variational equation:

$$\frac{\delta E_{KS}}{\delta \varphi_i^*(r)} = \frac{\delta T_S}{\delta \varphi_i^*(r)} + \left[\frac{\delta E_{ex}}{\delta n(r)} + \frac{\delta E_{Hartree}}{\delta n(r)} + \frac{\delta E_{xc}}{\delta n(r)}\right] \frac{\delta n(r)}{\delta \varphi_i^*(r)} = 0 \tag{19}$$

With the constraint of orthonormalization $\langle \varphi_i | \varphi_j \rangle = \delta_{i,j}$, this implies the form of Kohn Sham for Schrödinger equations:

$$(\hat{H}_{KS} - \varepsilon_i)\varphi_i(r) = 0 \tag{20}$$

ε_i represents the eigenvalues, and $\hat{H}_{KS}$ is the effective Hamiltonian H,

$$\hat{H}_{KS}(r) = -\frac{1}{2}\nabla^2 + V_{KS}(r) \tag{21}$$

$$V_{KS}(r) = V_{ex}(r) + \frac{\delta E_{Hartree}}{\delta n(r)} + \frac{\delta E_{xc}}{\delta n(r)} \tag{22}$$

Eqs. (20)–(22) are known equations of Kohn Sham, the density $n(r)$, and the resulting total energy E_{KS}. These equations are independent of any approximation on the functional $E_{XC}(n)$, resolution provides the exact values of the density and the energy of the ground state of the interacting system, provided that $E_{XC}(n)$ is known. The latter can be described in terms of Hohenberg Kohn function (12)

$$E_{xc}[n] = F_{HK}[n] - (T_S[n] + E_{Hartree}[n]) \tag{23}$$

$$E_{xc}[n] = \langle \hat{T} \rangle - T_S[n] + \langle \hat{V}_{\text{int}} \rangle - E_{Hartree}[n] \tag{24}$$

This energy is related to potential exchange-correlation $V_{xc} = \frac{\partial E_{xc}}{\partial n(r)}$.

For the exchange-correlation functional (LDA and GGA) and the parameter set of Bloch theorem and Brillouin zone, the content is detailed in our previous chapter [10].

4. Elastic coefficients of typical crystals solved by DFT method

4.1 Elastic strain energy formulas of cubic crystals

The matrix element of cubic crystal system is $\begin{bmatrix} c_{11} & c_{12} & c_{12} & 0 & 0 & 0 \\ c_{12} & c_{11} & c_{12} & 0 & 0 & 0 \\ c_{12} & c_{12} & c_{11} & 0 & 0 & 0 \\ 0 & 0 & 0 & c_{44} & 0 & 0 \\ 0 & 0 & 0 & 0 & c_{44} & 0 \\ 0 & 0 & 0 & 0 & 0 & c_{44} \end{bmatrix}$, where the subscript of matrix C corresponds to i and j. In the calculation of the elastic modulus, the strain component needs to be applied to the same δ.

Strain tensor exhibits a total of six independent components. When a strain is applied, the corresponding strain component e_i (i = 1~6) can be expressed as:

$$e = (e_1, e_2, e_3, e_4, e_5, e_6) \tag{25}$$

Submitting the cubic crystal matrix into Eqs. (6), we can obtain the following formula:

$$\Delta E = \frac{V}{2}(c_{11}e_1e_1 + c_{11}e_2e_2 + c_{11}e_3e_3 + c_{12}e_1e_2 + c_{12}e_1e_3 + c_{12}e_2e_1 + c_{12}e_2e_3 + c_{12}e_3e_1 + c_{12}e_3e_2 + c_{44}e_4e_4 + c_{44}e_5e_5 + c_{44}e_6e_6) \tag{26}$$

Firstly, considering the simplest constant C_{44}, i and j are equal to 4, and by the cubic matrix elements, $C_{44} = C_{55} = C_{66}$. By setting $e = (0, 0, 0, \delta, \delta, \delta)$ of Eq. (25) and submitting into Eq. (5), we get:

$$\Delta E = \frac{V_0}{2}(c_{44}e_4e_4 + c_{44}e_5e_5 + c_{44}e_6e_6) \tag{27}$$

Therefore,

$$\frac{\Delta E}{V_0} = \frac{3}{2}c_{44}\delta^2 \tag{28}$$

Similarly, setting $e = (\delta, \delta, 0, 0, 0, 0)$, we obtain:

$$\Delta E = \frac{V_0}{2}(c_{11}e_1e_1 + c_{11}e_2e_2 + c_{12}e_1e_2 + c_{12}e_2e_1) \tag{29}$$

Therefore,

$$\frac{\Delta E}{V_0} = (c_{11} + c_{12})\delta^2 \tag{30}$$

On this basis, if a strain $e = (\delta, \delta, \delta, 0, 0, 0)$ is imposed, we have:

$$\Delta E = \frac{V_0}{2}(c_{11}e_1e_1 + c_{11}e_2e_2 + c_{11}e_3e_3 + c_{12}e_1e_2 + c_{12}e_1e_3 + c_{12}e_2e_1 + c_{12}e_2e_3 + c_{12}e_3e_1 + c_{12}e_3e_2) \tag{31}$$

Therefore,

$$\frac{\Delta E}{V_0} = \frac{3}{2}(c_{11} + 2c_{12})\delta^2 \tag{32}$$

After fitting the three equations, we can obtain three independent elastic constants c_{11}, c_{12}, and c_{44}.

A transverse strain causes a change in shape without a change in volume. Thus, by applying these five specific strains mentioned above and selecting a series of δ strain amplitude, we can get the data from the ΔE-δ curve, respectively, and then obtain by fitting quadratic coefficient according to the corresponding applied strain methods mentioned in **Table 2**. Finally, we can get the independent elastic constants of cubic crystals by these simultaneous equations.

For the cubic phase, the criteria for mechanical stability are given by [11]:

$$C_{11} > 0, C_{44} > 0, C_{11} > |C_{12}|, (C_{11} + 2C_{12}) > 0 \tag{33}$$

4.2 Elastic strain energy formulas of hexagonal crystal structures

Matrix element of hexagonal crystal structure is as:

$$\begin{bmatrix} c_{11} & c_{12} & c_{13} & 0 & 0 & 0 \\ c_{12} & c_{11} & c_{13} & 0 & 0 & 0 \\ c_{13} & c_{13} & c_{33} & 0 & 0 & 0 \\ 0 & 0 & 0 & c_{44} & 0 & 0 \\ 0 & 0 & 0 & 0 & c_{44} & 0 \\ 0 & 0 & 0 & 0 & 0 & \frac{1}{2}(c_{11} - c_{12}) \end{bmatrix}$$

, system has a total of five independent components, of which independent elastic constants are c_{11}, c_{12}, c_{13}, c_{33}, and c_{44}.

Similarly, we can get the strain tensor expansion by strain energy, thus leading to the relationship between strain energy and second-order elastic coefficients of the hexagonal crystal:

Deformation tensor	LCEC
$e = (\delta, \delta, 0, 0, 0, 0)$	$2c_{11} + 2c_{12}$
$e = (\delta, -\delta, 0, 0, 0, 0)$	$2c_{11} - 2c_{12}$
$e = (0, 0, 0, 0, 0, \delta)$	$3c_{11} + 6c_{12}$
$e = (\delta, \delta, \delta, 0, 0, 0)$	$3c_{11} + 6c_{12}$
$e = (\delta^2/(1-\delta^2), 0, 0, 2\delta, 0, 0)$	$4c_{44}$
$e = (0, 0, 0, \delta, \delta, \delta)$	$3c_{44}$

Table 2.
Deformation tensors to calculate independent elastic constants of cubic crystal.

$$\rho_0 F(\eta_J, T) = \rho_0 F(\eta_J, T) + \frac{1}{2}\{c_{11}^T \eta_1\eta_1 + c_{11}^T \eta_2\eta_2 + c_{33}^T \eta_3\eta_3 + c_{44}^T \eta_4\eta_4 + c_{44}^T \eta_5\eta_5 + \frac{c_{11}^T - c_{12}^T}{2}\eta_6\eta_6 + 2c_{12}^T \eta_1\eta_2 + 2c_{13}^T \eta_1\eta_3 + 2c_{13}^T \eta_2\eta_3\Big\} + \dots + \frac{1}{n!}\sum_{JP\dots} c_{JP\dots}^T \eta_J\eta_P\dots \tag{34}$$

For hexagonal crystal, the basis vector of original cell can be taken as:

$$\vec{R} = \begin{pmatrix} a_1 \\ a_2 \\ a_3 \end{pmatrix} = \begin{pmatrix} \frac{\sqrt{3}}{2}a & \frac{1}{2}a & 0 \\ -\frac{\sqrt{3}}{2}a & \frac{1}{2}a & 0 \\ 0 & 0 & c \end{pmatrix} \tag{35}$$

where a and c are the lattice constants of crystals.
We can apply strain $e = (\delta, \delta, 0, 0, 0, 0)$ to calculate the value of $c_{11} + c_{12}$:

$$\frac{\Delta E}{V_0} = (c_{11} + c_{12})\delta^2 \tag{36}$$

Similarly, we can apply strain $e = (0, 0, 0, 0, 0, \delta)$ to calculate the value of $c_{11}-c_{12}$:

$$\frac{\Delta E}{V_0} = \frac{1}{4}(c_{11} - c_{12})\delta^2 \tag{37}$$

Also for c_{33}, we can apply strain $e = (0, 0, \delta, 0, 0, 0)$:

$$\frac{\Delta E}{V_0} = \frac{1}{2}c_{33}\delta^2 \tag{38}$$

Then, for c_{44}, we can apply strain $e = (0, 0, 0, \delta, \delta, 0)$:

$$\frac{\Delta E}{V_0} = c_{44}\delta^2 \tag{39}$$

Finally, we can apply strain $e = (\delta, \delta, \delta, 0, 0, 0)$ to calculate the values of c_{11}, c_{12}, c_{13}, and c_{33}:

$$\frac{\Delta E}{V_0} = \left(c_{11} + c_{12} + 2c_{13} + \frac{1}{2}c_{33}\right)\delta^2 \tag{40}$$

For hexagonal structure, mechanical stability condition is as [10]:

$$c_{44} > 0,\ c_{11} > |c_{12}|,\ (c_{11} + 2c_{12})c_{33} > 2c_{13}^2 \tag{41}$$

Similarly, by applying these five specific strains mentioned above and selecting a series of δ strain amplitude, the independent elastic constants of hexagonal crystals by these simultaneous equations of ΔE-δ data in **Table 3** can be obtained.

The corresponding applied strain methods are mentioned in **Table 2**. Elastic strain energy and strain relations for hexagonal structure can be given in **Table 3**.

Last but not the least, for elastic strain energy formulas of monoclinic crystals, the content is detailed in our previous chapter work [9].

Distortion matrix	Energy (Taylor's expansion)	$\frac{1}{V_0}\frac{\partial^2 E(V,\delta)}{\partial\delta^2}\Big\vert_{\delta=0}$
$\begin{pmatrix} 1+\delta & 0 & 0 \\ 0 & 1+\delta & 0 \\ 0 & 0 & 1 \end{pmatrix}$	$E(V,\delta) = E(V_0,0) + V_0\left[(\tau_1+\tau_2)\delta + (c_{11}+c_{12})\delta^2\right]$	$2(c_{11}+c_{12})$
$\begin{pmatrix} 1+\delta & 0 & 0 \\ 0 & 1-\delta & 0 \\ 0 & 0 & 1 \end{pmatrix}$	$E(V,\delta) = E(V_0,0) + V_0\left[(\tau_1-\tau_2)\delta + (c_{11}-c_{12})\delta^2\right]$	$2(c_{11}-c_{12})$
$\begin{pmatrix} 1 & 0 & 0 \\ 0 & 1 & 0 \\ 0 & 0 & 1+\delta \end{pmatrix}$	$E(V,\delta) = E(V_0,0) + V_0\left[\tau_3\delta + 1/2c_{33}\delta^2\right]$	c_{33}
$\begin{pmatrix} 1 & 0 & 0 \\ 0 & 1 & \delta \\ 0 & \delta & 1 \end{pmatrix}$	$E(V,\delta) = E(V_0,0) + V_0\left[\tau_5\delta + 2c_{44}\delta^2\right]$	$4c_{44}$
$\begin{pmatrix} 1+\delta & 0 & 0 \\ 0 & 1 & 0 \\ 0 & 0 & 1+\delta \end{pmatrix}$	$E(V,\delta) = E(V_0,0) + V_0\left[(\tau_1+\tau_3)\delta + 1/2(c_{11}+2c_{13}+c_{33})\delta^2\right]$	$c_{11}+2c_{13}+c_{33}$
$\begin{pmatrix} 1+\delta & 0 & 0 \\ 0 & 1 & 0 \\ 0 & 0 & 1-\delta \end{pmatrix}$	$E(V,\delta) = E(V_0,0) + V_0\left[(\tau_1+\tau_3)\delta + 1/2(c_{11}-2c_{13}+c_{33})\delta^2\right]$	$c_{11}-2c_{13}+c_{33}$
$\begin{pmatrix} 1+\delta & 0 & 0 \\ 0 & 1+\delta & 0 \\ 0 & 0 & 1+\delta \end{pmatrix}$	$E(V,\delta) = E(V_0,0) + V_0[(\tau_1+\tau_2+\tau_3)\delta + 1/2(2c_{11}+2c_{12}+4c_{13}+c_{33})\delta^2]$	$2c_{11}+2c_{12}+c_{33}+4c_{13}$

Table 3.
Deformation tensor to determine independent elastic constants of hexagonal crystal [12, 13].

5. Elastic coefficients of typical single crystals at nanoscale

5.1 Elastic coefficients C_{ij} of cubic crystals: CaO and MgO

All DFT calculations were performed by the total energy of the linear combination of atomic orbitals with a Hartree-Fock self-consistent field [14, 15] method implementation. Monkhorst Pack using the Brillouin zone integration method [16] is used to set K point to verify the convergence of different sampling meshes.

5.1.1 Modeling of cubic crystals: CaO and MgO

5.1.1.1 Lime (CaO) cubic structure

CaO is an alkaline earth oxide of native lime. In fact, there are many studies of CaO phase transition characteristics using quantum mechanics. The high-pressure structure phases and the elastic properties of CaO are important for mortar modeling in civil engineering. CaO has the B_1–B_2 transition at 61 GPa [17], and phase transformation from B_1 to B_2 takes over a range of pressure 60–70 GPa [18] from shockwave measurements. Kalpana et al. [19] and Baltache et al. [20] also investigated this transition for CaO. Mehl et al. [21] predicted elastic constants and their pressure dependences of CaO using potential-induced breathing (PIB) models. The ab initio pseudopotential calculations with plane-wave basis (PWPP) and ab initio full-potential linear muffin-tin-orbitals (FP-LMTO) are used to study high-pressure elastic properties of CaO by Karki and Crain [22] and Tsuchiya and Kawamura [23], respectively. Tsuchiya et al. [23] have used the full-potential linearized muffin-tin-orbital method to study elastic constants of MgO and CaO.

Lime (CaO) is a cubic structure, the initial lime crystal lattice by Fiquet et al. [24] is as: $a = b = c = 4.8330$ Å, $\alpha = \beta = \gamma = 90°$, with space group Fm3m.

As is shown in **Figure 2**, Ca and O are arranged alternately to occupy cube vertex, edge, and face-center. From the lattice unit terms, the ion at the apex, on the edge, on the surface, and inside of cell, each ion belonging to the cell separately have the occupancy rate of 1/8, 1/4, 1/2, and 1. Atomic coordinates and displacement parameters of lime (CaO) are given in **Table 4**.

The elastic properties of the B_1-phase CaO are calculated by using the plane-wave pseudopotential density functional theory (DFT) method. The Cambridge Serial Total Energy Package (CASTEP) program is used to obtain the calculated energy-volume points. The lattice constants, the bulk modulus, and the pressure derivatives of bulk modulus of the B_1-CaO at zero pressure are given. In addition, the elastic constants, shear modulus, and bulk modulus of the B_1-phase CaO are investigated over a wide pressure range from 0 to 1.0 GPa.

5.1.1.2 Periclase (MgO) cubic structure

In 1995, Duffy et al. [25] have found, for pressures up to 227 GPa, that B1-type MgO remained stable, while at 300 K, the stable pressure can reach up to 199 GPa. By Anvil hydrostatic test, Fei [26] has measured and got high-pressure state equation as well as elastic modulus of MgO. By ultrasonic interferometry method, Chen et al. [27] have found that the elastic anisotropy of MgO structure decreases with an increase of pressure under normal temperature conditions, while the elastic anisotropy increases with temperature at high pressure. Jaekson and Niesler [28] have obtained isothermal bulk modulus of MgO and its first derivative of pressure. Sinogeikin and Bass [29] have studied the pressure state equation of MgO by Anvil Brillouin scattering experiments. Zha et al. [30] have experimentally measured elastic properties with the pressure up to 55 GPa by scattering spectroscopy *X* Brillouin zone. Merkel et al. [31] have studied the pressure up to 47 GPa and measured elastic modulus, shear modulus, and elastic anisotropy. Wolf and Bukowinski [32] have investigated the stabilization of the ionic charge densities in MgO and CaO crystals by an electron gas model. Matsui et al. [33] have calculated elastic constants and thermodynamic properties of MgO structure over a wide

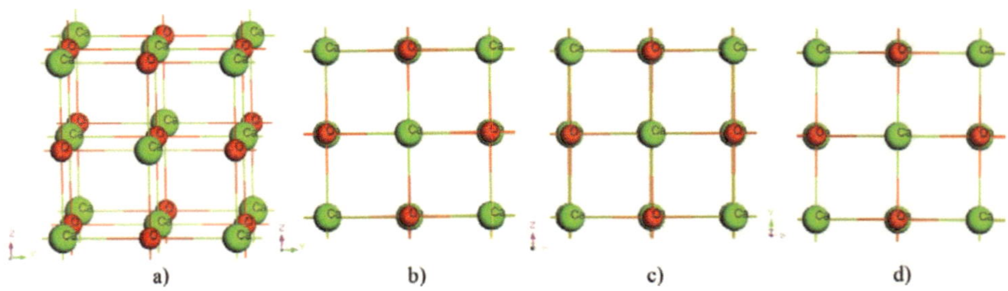

Figure 2.
Crystal structure of CaO. a) CaO cubic crystal b) crystal in x-direction c) crystal in y-direction d) crystal in z-direction.

Atom	x	y	z	Occupancy rate	Uiso or Ueq
Ca	0.000	0.000	0.000	1.00	1.00
O	0.500	0.500	0.500	1.00	1.00

Table 4.
Atomic coordinates and displacement parameters of lime (CaO) [24].

temperature and pressure range. Karki et al. [34], using the local density approximation of density functional perturbation theory, have studied the MgO thermal expansion coefficient and elastic properties. Karki et al. [35] have used a plane-wave qualitative study of the elastic properties of the potential B1-type MgO in the range of 0–25 GPa.

Periclase (MgO) is a cubic structure, the initial lime crystal lattice by Kalpana [19] is as: $a = b = c = 4.2130$ Å, $\alpha = \beta = \gamma = 90°$, with space group *Fm3m*.

As is shown in **Figure 3**, Mg and O are arranged alternately to occupy cube vertex, edge, and face-center. Mg^{2+} ions will fill into the O^{-2} ion accumulation to form MgO crystal.

Atomic coordinates and displacement parameters of MgO are given in **Table 5**.

Electronic basis is optimized in calculation using a linear combination of atomic orbitals by self-consistent field (SCF) method of Hartree Fock. The *K* points of convergence have been met when taking a $16 \times 16 \times 16$ mesh in our calculations. For complex space, we use parameter of Gilat and Raubenheimer [36]: when the total energy reaches less than 10^{-6} eV/atom, the self-consistent convergence is more effective.

In addition, the elastic constants, Young's modulus, and shear modulus are investigated over a wide pressure range from 0 to 1.0 GPa.

5.1.2 Initial conditions and elastic constants of CaO and MgO

5.1.2.1 Initial conditions and elastic constants of CaO

All the total energy electronic structure calculations are implemented through the CASTEP code. In CaO crystal modeling, we adopt the exchange-correlation interaction by using the GGA and LDA of the exchange-correlation function. Ultrasoft pseudopotentials with cutoff energy of the plane waves of 500 eV are used to describe the electron-ion interaction. For the Brillouin zone sampling, the $17 \times 17 \times 17$ of Monkhorst-Pack mesh [15] is used, where the self-consistent convergence of the total energy is at 10^{-6} eV per atom.

For CaO crystal at zero temperature and pressure equilibrium, lattice constant *a* and bulk modulus B_0 are as shown in **Table 6**.

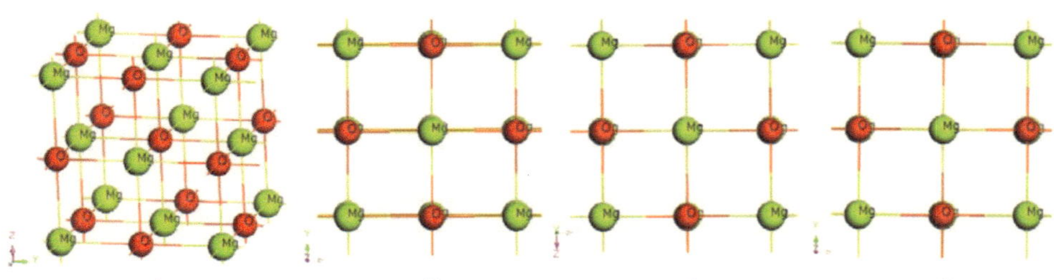

a) b) c) d)

Figure 3.
Crystal structure of MgO. (a) MgO cubic crystal. (b) Crystal in x-direction. (c) Crystal in y-direction. (d) Crystal in z-direction.

Atom	x	y	z	Occupancy rate	Uiso or Ueq
Mg	0.000	0.000	0.000	1.00	1.00
O	0.500	0.500	0.500	1.00	1.00

Table 5.
Atomic coordinates and displacement parameters of MgO [19].

	a	C_{11}	C_{12}	C_{44}	B	G
Present (LDA)	4.7124	289.497	71.432	76.652	144.12	88.295
Present (GGA)	4.8339	200.879	56.902	75.600	104.89	74.134
Ref. [21]	4.72	274	54	53	128	—
Ref. [22]	4.820	206	50	66	102	71
Ref. [23]	4.838	239	51.6	77.4	117	83.6
Ref. [24]	4.840	223	53	84	109	85
Exp. [37]		221.89	57.81	80.32		80.59
Exp. [38]		220.53	57.67	80.03		80.59

Table 6.
Lattice a, elastic constants, bulk and shear moduli of CaO at zero pressure.

The structural and the elastic properties of CaO are investigated by ab initio plane-wave pseudopotential density functional theory method. Lattice a, elastic constants, bulk and shear moduli of CaO under different pressures are shown in **Table** 7.

As can be seen from the **Table 8**, the obtained a, C_{ij}, B, and G of the B_1-phase CaO agree well with the experimental data [37, 38] and other calculated results [21–24]. CaO cubic crystal under pressure 0–1.0 GPa by GGA is shown in **Figure 4**.

From **Figure 4**, we can see the pressure dependences of the elastic constants c_{ij} of CaO at zero temperature and different pressures. It is found that c_{11} varies largely under the effect of pressure, compared with the variations in c_{12} and c_{44}. The change of elastic constant c_{11} represents a produced change with a longitudinal strain in length elasticity. The elastic constants c_{12} and c_{44} are related to the elasticity in shape, which is a shear constant. Besides, c_{12} and c_{44} are less sensitive of pressure as compared with c_{11}. As pressure increases, c_{11} and c_{12} increase monotonically, but c_{44} slightly decreases monotonically.

5.1.2.2 Initial conditions and elastic constants of MgO

Similarly, the exchange-correlation interaction by using the GGA and LDA of the exchange-correlation functions is adopted. Ultrasoft pseudopotentials with cut-off energy of the plane waves are set to 700 eV to describe the electron-ion

P	*a*	$V(Å^3)$	c_{11}	c_{12}	c_{44}
0.0 (LDA)	4.7124	104.647	289.497	71.432	76.652
0.0 (GGA)	4.8339	112.951	200.879	56.902	75.600
0.1 (GGA)	4.8324	112.847	201.548	56.742	75.621
0.2 (GGA)	4.8310	112.748	202.538	56.986	75.639
0.3 (GGA)	4.8299	112.665	203.062	56.893	75.659
0.4 (GGA)	4.8279	112.534	204.506	57.361	75.683
0.5 (GGA)	4.8264	112.428	205.667	57.736	75.717
0.6 (GGA)	4.8251	112.336	206.105	57.440	75.717
0.7 (GGA)	4.8238	112.249	207.001	57.664	75.753
0.8 (GGA)	4.8219	112.110	208.202	57.762	75.784
0.9 (GGA)	4.8204	112.008	209.415	58.192	75.890
1.0 (GGA)	4.8193	111.934	209.830	58.059	75.823

Table 7.
Lattice a, elastic constants, bulk and shear moduli of CaO under different pressures.

	a	C_{11}	C_{12}	C_{44}	*B*	*G*	*A*
Present (LDA)	4.2130	298.508	92.228	141.197	160.988	124.506	—
Present (GGA)	4.2130	343.468	106.358	140.823	185.395	131.449	—
Ref. [19]	4.2130	—	—	—	167.600	—	—
Ref. [22]	4.30	270	73	127	139	115	0.21
Ref. [24]	4.2590	318	87	144	164.000	132	0.18
Ref. [29]	—	298.96	96.42	157.13	163.930	131.80	—
Ref. [33]	4.2210	300	93.6	147	162.4	127.6	0.29
Ref. [35]	4.2510	291	91	139	158.00	121.800	0.27
Ref. [39]	4.16–4.24	286–352	91–108	158–188	157.0–198.0	—	—
Ref. [40]	4.212	—	—	—	—	—	—

Table 8.
Lattice a, elastic constants C_{ij}, bulk and shear moduli of MgO at zero pressure.

interaction. For Brillouin zone, the 15 × 15 × 15 of Monkhorst-Pack mesh [15] is used, where the self-consistent convergence of the total energy is at 10^{-6} eV per atom.

As can be seen in **Table 8**, the obtained a, C_{ij}, B, and G of the MgO agree well with the experimental data [29] and other calculated results [19, 22, 24, 33, 35, 39, 40]. It is noted that c_{12} and c_{44} for MgO are positive indicating that the MgO is stable at zero pressure (**Table 9**).

In the case of cubic crystals, there are only three independent elastic constants of c_{11}, c_{12}, and c_{44}. MgO cubic crystal under pressure 0–1.0 GPa by GGA is shown in **Figure 5**.

From **Figure 5**, we can see that c_{11} varies largely under the effect of pressure as compared with the variations in c_{12} and c_{44}.

5.2 Elastic coefficients C_{ij} of hexagonal crystals: CH and calcite

Portlandite is one of the main typical constituents of hydrated cementitious systems. CH represents 17–25% of the volume fraction of portland cement paste, and its Young's modulus is needed in the modeling of cement systems on the macro- and microscales [41, 42]. Portlandite influences the physical and mechanical properties as well as the durability of cement-based materials. Calcite has very large

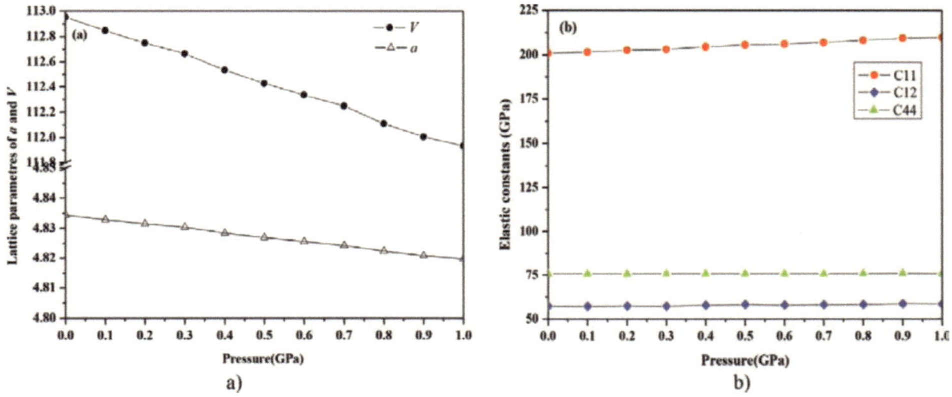

Figure 4.
CaO cubic crystal under pressure 0–1.0 GPa by DFT. (a) Relative change of V. (b) Elastic constants.

P	a	V (Å³)	c_{11}	c_{12}	c_{44}
0.0 (GGA)	4.2130	74.778	343.468	106.358	140.823
0.0 (LDA)	4.2130	74.778	298.508	92.228	141.197
0.1 (GGA)	4.2038	74.289	309.892	95.878	142.181
0.2 (GGA)	4.2030	74.246	308.989	94.200	142.178
0.3 (GGA)	4.2022	74.202	310.994	95.635	142.300
0.4 (GGA)	4.2012	74.153	312.885	96.635	142.371
0.5 (GGA)	4.2005	74.115	313.050	96.187	142.484
0.6 (GGA)	4.1996	74.069	313.565	96.092	142.519
0.7 (GGA)	4.1988	74.025	314.766	96.493	142.597
0.8 (GGA)	4.1979	73.980	314.612	95.599	142.677
0.9 (GGA)	4.1971	73.937	316.353	96.652	142.857
1.0 (GGA)	4.1963	73.893	317.067	96.738	142.940

Table 9.
Lattice a, elastic constants, bulk and shear moduli of MgO under different pressures.

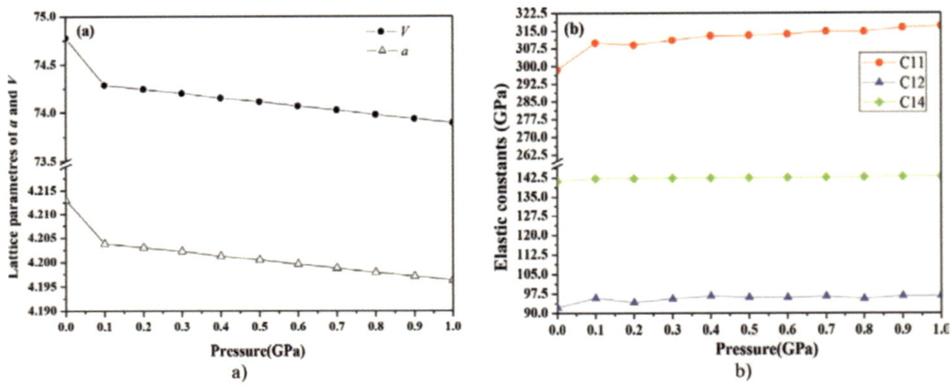

Figure 5.
MgO cubic crystal under pressure 0–1.0 GPa by DFT. (a) Relative change of V. (b) Elastic constants.

industrial applications in the field of civil engineering construction, and it can also be used as a flux in glass and in the field of metallurgy.

5.2.1 Modeling of hexagonal crystals: CH and calcite

5.2.1.1 Portlandite hexagonal structure

Laugesen [43] has calculated the elastic properties of portlandite using the density functional calculations, where elastic constants are given by GGA method. Speziale et al. [44] have determined the elastic constants of portlandite by Brillouin spectroscopy. Besides, Busing et al. [45] have investigated the calcium hydroxide structure by the neutron diffraction, of which the atomic information is given and thus is used to establish the CH model. The morphology of the portlandite crystal is hexagonal when the structure of the crystal is trigonal [45] with space group P3m1. Initial model of CH is as shown in **Figure 6**.

As shown in **Figure 6**, Ca^{2+} is octahedrally coordinated by oxygen. CH is a hexagonal structure, the initial crystal lattice [43] is: $a = b = 3.5930$ Å, $c = 4.9090$ Å, $\alpha = \beta = 90°$, $\gamma = 120°$. There is no hydrogen bond or other strong bonds across the layer. Ca^{2+} is octahedrally coordinated by oxygen and each O^{2-} is tetrahedrally

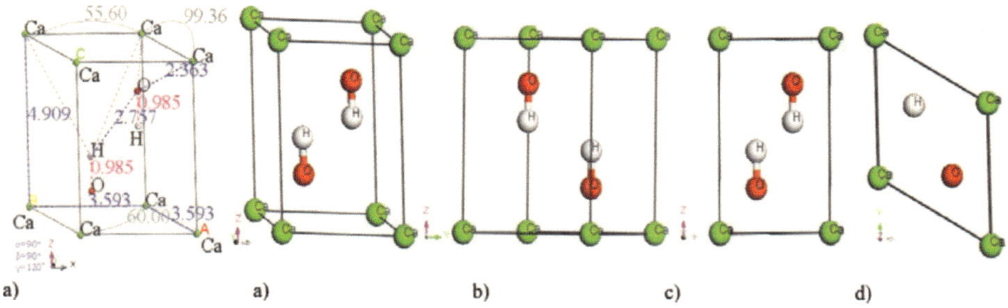

Figure 6.
Final optimized parameters of CH hexagonal crystal. (a) CH hexagonal crystal. (b) Real CH cell. (c) Cell in x-direction. (d) Cell in y-direction. (e) Cell in z-direction.

coordinated by calcium and hydrogen. The direction of hydroxyl groups formed by oxygen and hydrogen is vertical to the (001) plane direction. The atomic distance between O^{2-} and H^{+} is 2.757 Å (**Table 10**).

In order to determine whether the elastic coefficients under various pressures are stable, the pressure region of 0–5.0 GPa is added, with the same initial conditions of Laugesen [43].

5.2.1.2 Calcite hexagonal structure

Several researchers [46, 47] have been interested in elastic constants of calcite and its structure. Each carbon atom of calcite is the center of an equilateral triangle, and each vertex of the triangle is occupied by the oxygen atom. The calcite is crystallizing in the trigonal system "or rhombohedral" system. Zhang and Reeder [48] have investigated compressibilities of calcite-structure carbonates and confirmed the results found by Aydinol et al. [49] on the post- and pyroxene-aragonite phase type. Rohl et al. [50] have developed a new force field for calcite model. Effenberger et al. [51] discussed some aspects of the stereochemistry of calcite-type carbonates. Pilati et al. [52] have given the atomic displacement parameters in the calcite structure. Zaoui and Sekkal [53] have investigated the mechanisms behind the ikaite-to-calcite phase transformation by using molecular dynamics calculations. Le Page et al. [54] have investigated calcite using ab initio calculation to study their stiffness properties. Meanwhile, the elastic properties of calcite structure have been studied by Chen et al. [55] using Brillouin spectroscopy. Dandekar and Ruoff [56] have investigated the MgO structure on the temperature dependence of the elastic constants. Prencipe et al. [57] have studied its structure and properties using an ab initio quantum-mechanical calculation.

Initial hexagonal model of calcite by Prencipe [57] is as: $a = b = 5.0492$ Å, $c = 17.343$ Å, $\alpha = \beta = 90°$, $\gamma = 120°$, with space group R-3C. The calcite model is as shown in **Figure 7**.

The initial lattice of calcite hexagonal crystal is as shown in **Figure 7(a)**. From **Figure 7(a)** and **(b)**, calcite crystal has the following features: (1) two Ca atoms are connected to the anionic groups (CO_3^{-2}), form together an equilateral triangle, including the central carbon atom and each vertex totalizing three oxygen atoms.

Atomic species	X	Y	Z	Occupancy rate	Uiso or Ueq
Ca	0.0000	0.0000	0.0000	1.00	1.00
O	0.3333	0.6667	0.2346	1.00	1.00
H	0.3333	0.6667	0.4280	1.00	1.00

Table 10.
Atomic coordinates and displacement parameters of portlandite [45].

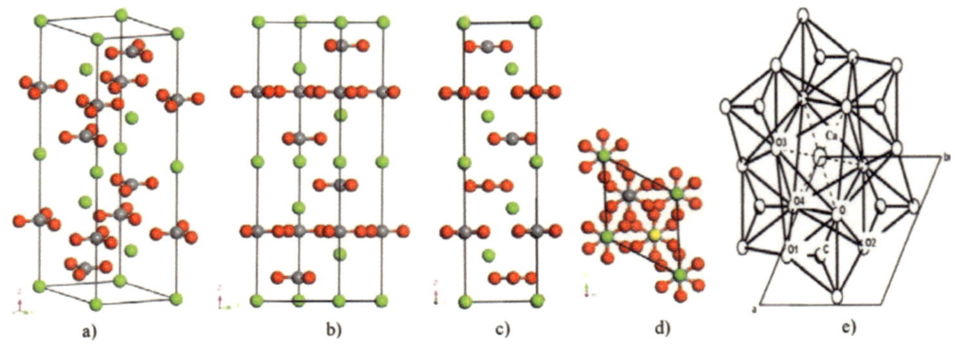

Figure 7.
Calcite hexagonal model. (a) Calcite hexagonal model. (b) In x-direction. (c) In y-direction. (d) In z-direction. (e) CO_3 group.

Atomic species	X	Y	Z	Occupancy rate	Uiso or Ueq
Ca	0.0000	0.0000	0.0000	1.00	1.00
C	0.0000	0.0000	0.2500	1.00	1.00
O	0.25642	0.0000	0.2500	1.00	1.00

Table 11.
Atomic coordinates and displacement parameters of calcite [57].

(2) CO_3 group interatomic distances are as: C—O is 1.2815 Å, Ca—O_1 is 3.4589 Å, O—O_1 is 2.2197 Å, O—O_2 is 4.7911 Å, O—O_3 is 3.4107 Å, and O—O_4 is 4.2459 Å (**Table 11**).

In order to determine whether the elastic coefficients under various pressures are stable, the pressure region of 0–5.0 GPa using DFT methods is added.

5.2.2 Initial conditions and elastic constants of CH and calcite

5.2.2.1 Initial conditions and elastic constants of CH

Initial conditions are as: *GGA* of the exchange-correlation function is used. Ultrasoft pseudopotentials with cutoff energy of the plane waves are 700 eV. Brillouin zone is 12 × 12 × 12. The self-consistent convergence of the total energy is

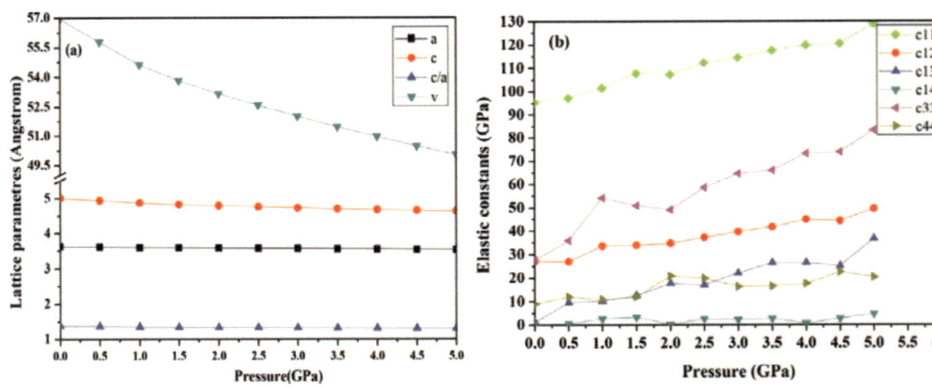

Figure 8.
CH hexagonal crystal under pressure 0–5.0 GPa by DFT. (a) Relative change of a, c, and V. (b) Elastic constants.

	Pressure	a_0(Å)	c_0(Å)	V_0 (Å^3)	C_{11}	C_{12}	C_{13}	C_{14}	C_{33}	C_{44}	C_{66}
Cal. [43]	0	3.5930	4.9090	54.8830	99.39	30.78	7.36	0	36.29	7.88	34.305
Exp. [44]	0	3.5927 (±0.0009)	4.908 (±0.005)	54.87 (±0.06)	102.0 (±2.0)	32.0 (±1.0)	8.4 (±0.4)	4.5 (±0.2)	33.6 (±0.7)	12.0 (±0.3)	35.0 (±1.1)
Cal. [58]	0	3.6580	4.7930	55.5000	135.1	45.6	6.6	0	26.0	4.1	44.7
Exp. [59]	0	3.5890	4.9110	54.8000	99.28 (±1.5)	36.21 (±2.0)	29.65 (±15.0)	≈0 (assumed)	32.6 (±2.0)	9.864 (±2.0)	31.55 (±1.5)
Present	0.0 (GGA)	3.5918	4.9063	54.8162	103.948	30.923	2.564	4.336	42.670	14.140	36.513
	0.0 (LDA)	3.6233	5.0086	56.9445	95.417	27.162	1.313	0.242	27.585	8.998	34.127
	0.5 (GGA)	3.6118	4.9383	55.7899	97.140	27.042	9.553	0.503	35.899	12.003	35.049
	1.0 (GGA)	3.5999	4.86716	54.6274	101.463	33.613	10.193	2.725	54.182	10.904	33.925
	1.5 (GGA)	3.5903	4.8207	53.8137	107.685	33.889	12.593	3.396	50.867	12.045	36.898
	2.0 (GGA)	3.5815	4.7855	53.1590	107.236	34.681	17.650	0.285	49.072	20.818	36.278
	2.5 (GGA)	3.5733	4.755	52.5737	112.249	37.190	17.139	2.661	58.666	19.959	37.529
	3.0 (GGA)	3.5651	4.7245	52.0042	114.388	39.664	22.111	2.372	64.582	16.340	37.362
	3.5 (GGA)	3.5572	4.6961	51.4622	117.505	41.661	26.485	2.846	65.960	16.476	37.922
	4.0 (GGA)	3.5492	4.6718	50.9649	119.696	44.869	26.498	0.648	73.285	17.543	37.414
	4.5 (GGA)	3.5415	4.6461	50.4651	120.400	44.261	25.089	2.851	73.884	22.543	38.070
	5.0 (GGA)	3.5342	4.6258	50.0389	128.865	49.445	36.702	4.781	83.203	20.346	39.710

Table 12.
Elastic constants (GPa) of CH under various pressures.

at 10^{-6} eV per atom. The simulation of CH hexagonal crystal under pressure 0–5.0 GPa by GGA method is shown in **Figure 8**.

The changes of elastic constants (GPa) of CH under various pressures are separately calculated, and the results are shown in **Table 12**.

From **Figure 8**, for CH crystal at 0 GPa, elastic coefficients under initial conditions by DFT are as: c_{11} = 103.948 GPa, c_{12} = 30.923 GPa, c_{13} = 2.564 GPa, c_{33} = 42.670 GPa, and c_{44} = 14.140 GPa. C_{14} is turned out not to be zero, which is different to the result of Laugesen. We can see that all elastic coefficients (c_{14} not included) under various pressures are stable except for a pressure of 1.0 and 4.5 GPa.

5.2.2.2 Initial conditions and elastic constants of calcite

Generalized gradient approximation (*GGA*) of the exchange-correlation function is used. Ultrasoft pseudopotentials with cutoff energy of the plane waves are 700 eV. Brillouin zone is 12 × 12 × 12. The self-consistent convergence of the total energy is 10^{-6} eV per atom.

Results of hexagonal calcite crystal under pressure 0–5.0 GPa by GGA is shown in **Figure 9**.

From **Figure 9**, for calcite crystal, elastic coefficients under initial conditions at 0 GPa by DFT are given as follows: c_{11} = 143.938 GPa, c_{12} = 55.579 GPa, c_{13} = 50.970 GPa, c_{33} = 81.175 GPa, and c_{44} = 32.658 GPa. The elastic coefficients of calcite under pressure 0–5 GPa show slight growing trends except for c_{11} and c_{12} within a pressure region of 2.5–3.5 GPa.

The calculated elastic constants of calcite are given in **Table 13**.

From **Table 13**, elastic constants have been calculated by GGA method as follows: c_{11} = 143.938 GPa, c_{12} = 55.579 GPa, c_{13} = 50.970 Pa, c_{33} = 81.175 GPa, and c_{44} = 32.658 GPa. By LDA method, elastic constants have been calculated as follows: c_{11} = 153.926 GPa, c_{12} = 55.806 GPa, c_{13} = 50.970 Pa, c_{33} = 82.702 GPa, and c_{44} = 59.991 GPa.

Similarly, for elastic constants C_{ij} of monoclinic crystals (gypsum and 11 Å tobermorite), the content is detailed in our previous chapter work [9]. Furthermore, elastic moduli of cubic, hexagonal, and monoclinic structures at larger scale can be homogenized by the calculated elastic constants mentioned, as in Chapter 4.

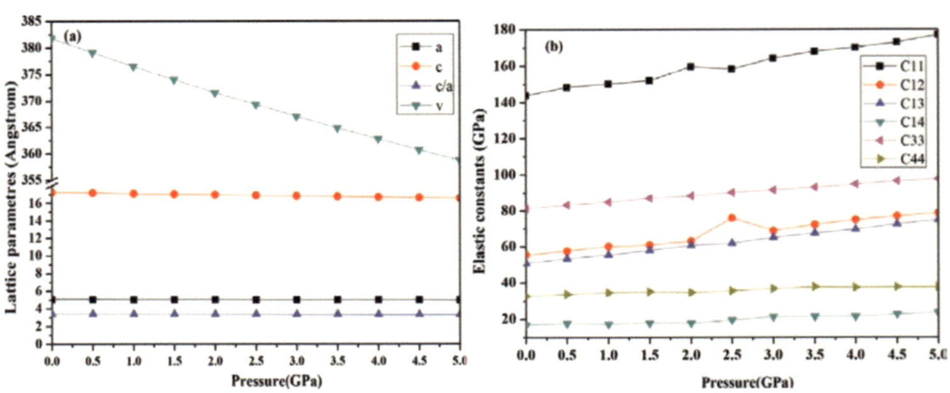

Figure 9.
Calcite hexagonal crystal under pressure 0–5.0 GPa by DFT. (a) Relative change of a, c, and V. (b) Elastic constants.

P (GPa)	a_0 (Å)	c_0 (Å)	c_0/a_0	V_0 (Å^3)	c_{11}	c_{12}	c_{13}	c_{14}	c_{33}	c_{44}	c_{66}
Ref. [46]	—	—	—	—	146.26	59.70	50.76	−20.76	85.31	34.05	43.28
Ref. [48]	4.989	17.044	—	367.39	—	—	—	—	—	—	—
Ref. [49]	4.981	15.902	—	341.676	—	—	—	—	—	—	—
Ref. [51]	4.990	17.061	—	367.8	—	—	—	—	—	—	—
Ref. [54]	4.988	17.068	—	367.761	159	70	63	19.5	98	39	44.5
Ref. [55]	—	—	—	—	149.4	57.9	53.5	20	85.2	34.1	45.75
Ref. [56]	—	—	—	—	145.7	55.9	53.5	20.5	85.3	33.4	44.9
Ref. [8]	5.061	17.097	—	379.279	152.3	57.05	54.83	17.14	87.77	36.00	47.625
0.0 (LDA)	4.9718	16.5296	3.3247	353.526	153.9261	55.8064	50.9698	13.6541	82.7015	33.9451	59.9905
0.0 (GGA)	5.0548	17.2551	3.4136	381.818	143.9383	55.5791	50.9698	16.8678	81.1753	32.6583	44.1796
0.5 (GGA)	5.0478	17.1808	3.4036	379.127	148.479	57.7574	53.4967	17.6855	83.2565	33.6887	45.3608
1.0 (GGA)	5.0411	17.1089	3.3939	376.536	150.1629	60.2232	55.5284	17.3573	84.8005	34.5971	44.9699
1.5 (GGA)	5.0357	17.0338	3.3826	374.074	152.0584	60.9377	58.008	17.8412	87.0412	35.0652	45.5604
2.0 (GGA)	5.0293	16.9642	3.3731	371.609	159.5886	63.1081	60.7533	17.8817	88.253	34.6942	48.2403
2.5 (GGA)	5.0245	16.8939	3.3623	369.350	158.2239	75.896	61.9659	19.5687	90.0313	35.5319	41.1640
3.0 (GGA)	5.0194	16.8242	3.3518	367.090	164.1449	68.8673	65.1085	21.3005	91.4266	36.7669	47.6388
3.5 (GGA)	5.0144	16.7569	3.3418	364.885	167.8984	72.1265	67.4523	21.4184	92.9625	37.7036	47.8860
4.0 (GGA)	5.0101	16.6887	3.3310	362.789	170.1009	74.8428	69.618	21.4534	94.683	37.3224	47.6291
4.5 (GGA)	5.0049	16.6285	3.3224	360.724	172.8895	76.9844	72.4475	22.8814	96.4539	37.6871	47.9526
5.0 (GGA)	5.0009	16.5620	3.3118	358.704	177.1157	78.6186	74.8454	23.6571	97.6551	37.548	49.2486

Table 13.
The calculated elastic constants of calcite (unit: GPa).

6. Conclusions

Elastic constants of typical crystal structures under a certain pressure region are calculated by DFT method. Three types of the cubic, hexagonal, and monoclinic crystals are considered, which have a certain value for both application and reference. Results are as:

1) For cubic structure, results are as:

(*i*) Elastic constants of CaO single crystal are as: c_{11} = 200.879 GPa, c_{12} = 56.902 GPa, and c_{44} = 75.600 GPa.

(*ii*) Elastic constants of our MgO single crystal are as: c_{11} = 298.508 GPa, c_{12} = 92.228 GPa, and c_{44} = 141.197 GPa.

2) For hexagonal structure, results are as:

(*i*) For CH crystal, elastic constants under initial conditions at 0 GPa based on GGA method are as: c_{11} = 103.948 GPa, c_{12} = 30.923 GPa, c_{13} = 2.564 GPa, c_{14} = 4.336 GPa, c_{33} = 42.670 GPa, and c_{44} = 14.140 GPa. Besides, elastic constants based on LDA method are as: c_{11} = 95.417 GPa, c_{12} = 27.162 GPa, c_{13} = 1.313 GPa, c_{14} = 0.242 GPa, c_{33} = 27.585 GPa, and c_{44} = 8.998 GPa. We can see that C_{14} is not zero, which is different to the result of Laugesen. The elastic coefficients under various pressures are stable except for a pressure of 0.8 GPa.

(*ii*) For calcite crystal, elastic coefficients under initial conditions at 0 GPa by DFT are given as follows: c_{11} = 143.938 GPa, c_{12} = 55.579 GPa, c_{13} = 50.970 GPa, c_{33} = 81.175 GPa, and c_{44} = 32.658 GPa. The elastic coefficients of calcite under pressure 0–5 GPa show slight growing trends except for a pressure of 2.5–3.5 GPa.

Moreover, for monoclinic single-crystal structure (11 Å tobermorite and gypsum), elastic coefficients are detailed in our previous chapter work of "Elastic constants and homogenized moduli of monoclinic structures based on density functional theory."

Acknowledgements

The author acknowledges the financial support provided by the China Scholarship Council and the startup foundation of Xi'an Shiyou University. Thanks Qiufeng Wang for her proofreading, and very sincere thankness to prof. Fabrice Bernard and prof. Siham Kamali-Bernard for the their enthusiastic and responsible guideness.

Author details

Jia Fu[1,2]

1 Xi'an Shiyou University, Xi'an, China

2 INSA de Rennes, Rennes, France

*Address all correspondence to: fujia@xsyu.edu.cn

References

[1] Messaoudi IS, Zaoui A, Ferhat M. Band-gap and phonon distribution in alkali halides. Physica Status Solidi B. 2015;**252**(3):490-495

[2] Thomas LH. The calculation of atomic fields. Mathematical Proceedings of the Cambridge Philosophical Society. Cambridge University Press. 1927; **23**(05):542-548

[3] Dirac PAM. Note on exchange phenomena in the Thomas-Fermi atom. Mathematical Proceedings of the Cambridge Philosophical Society. 1930; **26**:376-385

[4] Kohn W, Sham LJ. Self-consistent equations including exchange and correlation effects. Physical Review. 1965;**140**(4A):A1133

[5] Levy M, Perdew JP, Sahni V. Exact differential equation for the density and ionization energy of a many-particle system. Physical Review A. 1984;**30**(5): 2745

[6] Foulkes WMC, Mitas L, Needs RJ, et al. Quantum Monte Carlo simulations of solids. Reviews of Modern Physics. 2001;**73**(1):33

[7] Aulbur WG, Jönsson L, Wilkins JW. Quasiparticle calculations in solids. Solid State Physics. 1999;**54**:1-218

[8] Aaeid A. Études ab initio et dynamique moléculaire des propriétés structurales et thermodynamiques de la calcite et la witherite sous hautes pressions [PhD Thesis]. France: l'Université des Sciences et Technologies de Lille; 2010 (In French)

[9] Born M, Oppenheimer R. Zur quantentheorie der molekeln. Annalen der Physik. 1927;**389**(20):457-484

[10] Jia F. Elastic constants and homogenized moduli of monoclinic structures based on density functional theory. In: Yang G, editor. Density Functional Calculations: Recent Progresses of Theory and Application. Stafa-Zurich: Trans Tech Publications; 2018

[11] Wu ZJ, Zhao EJ, Xiang HP, et al. Crystal structures and elastic properties of superhard IrN2 and IrN3 from first principles. Physical Review B. 2007; **76**(5):054115

[12] Liu ZL, Chen XR, Wang YL. First-principles calculations of elastic properties of LiBC. Physica B: Condensed Matter. 2006;**381**(1):139-143

[13] Panda KB, Chandran KR. Determination of elastic constants of titanium diboride (TiB_2) from first principles using FLAPW implementation of the density functional theory. Computational Materials Science. 2006;**35**(2):134-150

[14] Pisani C, Dovesi R. Exact-exchange Hartree-Fock calculations for periodic systems. I. Illustration of the method. International Journal of Quantum Chemistry. 1980;**17**(3):501-516

[15] Pisani C, Dovesi R, Roetti C. Hartree-Fock ab initio treatment of crystalline systems. Berlin: Springer Science & Business Media; 2012

[16] Monkhorst HJ, Pack JD. Special points for Brillouin-zone integrations. Physical Review B. 1976;**13**(12):5188

[17] Jeanloz R, Ahrens TJ, Mao HK, et al. B1-B2 transition in calcium oxide from shock-wave and diamond-cell experiments. Science. 1979;**206**(4420): 829-830

[18] Mammone JF, Mao HK, Bell PM. Equations of state of CaO under static pressure conditions. Geophysical Research Letters. 1981;**8**(2):140-142

[19] Kalpana G, Palanivel B, Rajagopalan M. Electronic and structural properties of alkaline-earth oxides under high pressure. Physical Review B. 1995; **52**(1):4

[20] Baltache H, Khenata R, Sahnoun M, et al. Full potential calculation of structural, electronic and elastic properties of alkaline earth oxides MgO, CaO and SrO. Physica B: Condensed Matter. 2004;**344**(1):334-342

[21] Mehl MJ, Hemley RJ, Boyer LL. Potential-induced breathing model for the elastic moduli and high-pressure behavior of the cubic alkaline-earth oxides. Physical Review B. 1986;**33**(12): 8685

[22] Karki BB, Crain J. Structure and elasticity of CaO at high pressure. Journal of Geophysical Research: Solid Earth (1978–2012). 1998;**103**(B6): 12405-12411

[23] Tsuchiya T, Kawamura K. Systematics of elasticity: Ab initio study in B1-type alkaline earth oxides. Journal of Chemical Physics. 2001;**114**(22): 10086-10093

[24] Fiquet G, Richet P, Montagnac G. High-temperature thermal expansion of lime, periclase, corundum and spinel. Physics and Chemistry of Minerals. 1999;**27**(2):103-111

[25] Duffy TS, Hemley RJ, Mao H. Equation of state and shear strength at multimegabar pressures: Magnesium oxide to 227 GPa. Physical Review Letters. 1995;**74**(8):1371

[26] Fei Y. Effects of temperature and composition on the bulk modulus of (Mg, Fe) O. American Mineralogist. 1999;**84**(3):272-276

[27] Chen G, Liebermann RC, Weidner DJ. Elasticity of single-crystal MgO to 8 gigapascals and 1600 kelvin. Science. 1998;**280**(5371):1913-1916

[28] Jackson, Niesler H. High-pressure research in geophysics. In: Akimoto S, Manghnani MH, editors. Series: Advances in Earth and Planetary Sciences. Vol. 12. 1982. p. P93

[29] Sinogeikin SV, Bass JD. Single-crystal elasticity of pyrope and MgO to 20 GPa by Brillouin scattering in the diamond cell. Physics of the Earth and Planetary Interiors. 2000;**120**(1):43-62

[30] Zha CS, Mao H, Hemley RJ. Elasticity of MgO and a primary pressure scale to 55 GPa. Proceedings of the National Academy of Sciences. 2000;**97**(25):13494-13499

[31] Merkel S, Wenk HR, Shu J, et al. Deformation of polycrystalline MgO at pressures of the lower mantle. Journal of Geophysical Research: Solid Earth (1978–2012). 2002;**107**(B11):2271

[32] Wolf GH, Bukowinski MST. Variational stabilization of the ionic charge densities in the electron-gas theory of crystals: Applications to MgO and CaO. Physics and Chemistry of Minerals. 1988;**15**(3):209-220

[33] Matsui M, Parker SC, Leslie M. The MD simulation of the equation of state of MgO: Application as a pressure calibration standard at high temperature and high pressure. American Mineralogist. 2000;**85**(2):312-316

[34] Karki BB, Wentzcovitch RM, De Gironcoli S, et al. High-pressure lattice dynamics and thermoelasticity of MgO. Physical Review B. 2000;**61**(13):8793

[35] Karki BB, Stixrude L, Clark SJ, et al. Structure and elasticity of MgO at high pressure. American Mineralogist. 1997; **82**(1):51-60

[36] Gilat G, Raubenheimer LJ. Accurate numerical method for calculating frequency-distribution functions in solids. Physical Review. 1966;**144**(2): 390-395

[37] Chang ZP, Graham EK. Elastic properties of oxides in the NaCl-structure. Journal of Physics and Chemistry of Solids. 1977;**38**(12): 1355-1362

[38] Oda H, Anderson OL, Isaak DG, Suzuki I. Measurement of elastic properties of single-crystal CaO up to 1200 K. Physics and Chemistry of Minerals. 1992;**19**(2):96-105

[39] Marinelli F, Lichanot A. Elastic constants and electronic structure of alkaline-earth chalcogenides. Performances of various hamiltonians. Chemical Physics Letters. 2003;**367**(3): 430-438

[40] Speziale S, Zha CS, Duffy TS, Hemley RJ, Mao HK. Quasi-hydrostatic compression of magnesium oxide to 52 GPa—Implications for the pressure-volume-temperature equation of state. Journal of Geophysical Research. 2001; **106**(B1):515-528

[41] Kamali-Bernard S, Bernard F. Effect of tensile cracking on diffusivity of mortar: 3D numerical modelling. Computational Materials Science. 2009; **47**:178-185

[42] Bernard F, Kamali-Bernard S, Prince W. 3D multi-scale modelling of mechanical behavior of sound and leached mortar. Cement and Concrete Research. 2008;**38**:449-458

[43] Laugesen JL. Density functional calculations of elastic properties of portlandite, $Ca(OH)_2$. Cement and Concrete Research. 2005;**35**(2):199-202

[44] Speziale S, Reichmann HJ, Schilling FR, Wenk HR, et al. Determination of the elastic constants of portlandite by Brillouin spectroscopy. Cement and Concrete Research. 2008;**38**(10): 1148-1153

[45] Busing WR, Levy HA. Neutron diffraction study of calcium hydroxide. The Journal of Chemical Physics. 1957; **26**(3):563-568

[46] Dandekar DP. Pressure dependence of the elastic constants of calcite. Physical Review. 1968;**172**(3):873-877

[47] Wang Q. A computational study of calcium carbonate [Doctoral dissertation]. London: UCL, University College London; 2011

[48] Zhang J, Reeder RJ. Comparative compressibilities of calcite-structure carbonates: Deviations from empirical relations. American Mineralogist. 1999; **84**:861-870

[49] Aydinol MK, Mantese JV, Alpay SP. A comparative ab initio study of the ferroelectric behaviour in KNO_3 and $CaCO_3$. Journal of Physics: Condensed Matter. 2007;**19**(49):496210

[50] Rohl AL, Wright K, Gale JD. Letters. Evidence from surface phonons for the (2×1) reconstruction of the $(10\bar{1}4)$ surface of calcite from computer simulation. American Mineralogist. 2003;**88**(5–6):921-925

[51] Effenberger H, Mereiter K, Zemann J. Crystal structure refinements of magnesite, calcite, rhodochrosite, siderite, smithonite, and dolomite, with discussion of some aspects of the stereochemistry of calcite type carbonates. Zeitschrift für Kristallographie – Crystalline Materials. 1981;**156**(1–4):233-244

[52] Pilati T, Demartin F, Gramaccioli CM. Lattice-dynamical estimation of atomic displacement parameters in carbonates: Calcite and aragonite $CaCO_3$, dolomite CaMg $(CO_3)_2$ and magnesite $MgCO_3$. Acta Crystallographica Section B: Structural Science. 1998;**54**(5):515-523

[53] Zaoui A, Sekkal W. Mechanisms behind the ikaite-to-calcite phase transformation from molecular

dynamics calculations. Geoderma. 2014; **235**:329-333

[54] Le Page Y, Saxe P, Rodgers JR. Ab initio stiffness for low quartz and calcite. Physica Status Solidi B. 2002; **229**(3):1155-1161

[55] Chen CC, Lin CC, Liu LG, et al. Letters. Elasticity of single-crystal calcite and rhodochrosite by Brillouin spectroscopy. American Mineralogist. 2001;**86**(11–12):1525-1529

[56] Dandekar DP, Ruoff AL. Temperature dependence of the elastic constants of calcite between 160 and 300 K. Journal of Applied Physics. 1968; **39**(13):6004-6009

[57] Prencipe M, Pascale F, Zicovich-Wilson CM, Saunders VR, Orlando R, Dovesi R. The vibrational spectrum of calcite ($CaCO_3$): An ab initio quantum-mechanical calculation. Physics and Chemistry of Minerals. 2004;**31**(8): 559-564

[58] Kerisit S, Parker SC, Harding JH. Atomistic simulation of the dissociative adsorption of water on calcite surfaces. The Journal of Physical Chemistry B. 2003;**107**(31):7676-7682

[59] Holuj F, Drozdowski M, Czajkowski M. Brillouin spectrum of $Ca(OH)_2$. Solid State Communications. 1985;**56**(12): 1019-1021

Chapter 4

Homogenized Moduli of Typical Polycrystals by Elastic Constants Based on DFT

Jia Fu

Abstract

X-ray method to test the material properties and to obtain elastic constants is commonly based on the Reuss and Kroner models. Y parameter has been turned out to be an effective method to estimate elastic properties of polycrystalline material. Since Y parameters of cubic polycrystalline material based on the certain uniform stress (Reuss model) have not been given, our work aims to complete this part of the theoretical analysis which can effectively compare elastic constants measured by the X-ray diffraction method. The structural and the elastic properties of cubic structures (CaO and MgO) and hexagonal structures (CH and Calcite $CaCO_3$) are investigated by the density functional theory method, and then, the credibility of Y parameters for determining elastic moduli of cubic structures is proved and elastic properties in typical crystallographic planes of [100], [110], and [111] are also calculated. Meanwhile, Young's moduli of CaO, MgO, CH, and calcite structure are 175.76, 293.17, 58.08, 84.549, 54.30, and 79.51 GPa, which are all close to references. Elastic properties of cubic and hexagonal structures under various pressures are calculated and the surface constructions of elastic moduli are drawn, showing the anisotropy at various directions. Then, the bulk modulus, shear modulus, Young's modulus, Poisson's ratio, and anisotropy factor are determined.

Keywords: anisotropic elasticity, Young's modulus, Y parameter, cubic crystal, hexagonal crystal, polycrystalline materials

1. Introduction

The development of computational methods to describe and predict the mechanical properties of materials is of obvious practical importance. The accuracy of the prediction of mechanical or, from a more general point of view, physical properties depends largely on the knowledge of the intrinsic properties of the various constitutive phases. Up to now, these properties have been solely obtained by experimental procedures. The tremendous increase of computational capabilities has largely favored the development of numerical modeling based on a realistic and multiscale description of these kinds of materials. Since several years, some of the authors of this chapter have been involved in the development of such numerical tools from the micro- to the macrolevel (see, for instance, [1] or [2]). The intent is to assist the material developer by providing a rational approach to material development and concurrently assist the structural designer by providing an integrated analysis tool that incorporates fundamental material behavior. It is now clear that any realistic attempts to accomplish such an objective should be based on appropriate nanostructure/performance relationships, and also on an efficient method to upscale the properties determined on the lowest level to the next higher one. This is the context of the work detailed in this chapter.

Generally, many industrial and engineering materials consisted of dissimilar constituents are essentially inhomogeneous, which are distinguishable at a small length scale. Each constituent composed by anisotropic crystals shows different material properties and/or material orientations at some smaller length scales. At a sufficiently fine scale, all materials can be considered to be heterogeneous. The crystal anisotropy would be very difficult to handle computationally when the crystals are overlapping, the structure of isotropic polycrystals as the real materials at large scale are usually isotropic. The length-scale problem is considered as a great importance factor in material mechanics. Each material has structural features from the microstructure when studied at sufficiently physical parameters, from the view of homogenization. Polycrystalline structure contains a variety of information (e.g., orientation) and the properties of a single crystal, such as anisotropy. Within the mechanics of typical crystal structures, the transition from the micro- to the mesoscale (homogenization) and vice versa (localization) can be estimated.

Homogenization is an idealized description of a statistical distribution inside the actual heterogeneous material. At the engineering level, there exists such a scale (intermediate between the microscopic scale—that of the constituents—and the scale of the structure); it is said that the material can be homogenized. As opposed to the scale of constituents, the macroscopic homogeneity is called to be of statistical homogeneity or microscopic. The concept of homogenization is gradually proposed, the Gologanu-Leblond-Devaux's (GLD) analysis of an rigid-ideal plastic (von Mises) spheroidal volume gradually is extended to the case when the matrix is anisotropic (obeying Hill's anisotropic yield criterion), and the representative volume element (RVE) is subjected to arbitrary deformation. For quasibrittle materials, Zhu et al. [3] have formulated the anisotropic model in the framework of Eshelby-based homogenization methods. However, for the homogenization of elastic deformation, especially for polycrystalline structures, it is still the traditional Reuss-Voigt-Hill method. The homogenized moduli can be determined using the homogenization theory where periodic media are implemented. The hypothesis of continuity of a material is introduced to overcome theoretical difficulty, which implies a notion of statistical average, and the actual constitution of the material is idealized by considering the material to be continuous. Once the continuity model is admitted, the concept of homogeneity is deduced from it [4]. A homogeneous medium is then characterized by properties that are identical at every point. This hypothesis is particularly suitable for polycrystalline model.

1.1 Limits of X-ray diffraction methods and Voigt-Reuss-Hill averaging

Macroscopic averaging assuming a random distribution of initial crystals leads to isotropic mechanical properties such as Young's modulus. The upscaling to higher levels is then done by means of the Y-parameter theory which has been introduced by Zheng et al. [5, 6] a few years ago to overcome the limits of X-ray diffraction methods and conventional Voigt-Reuss-Hill averaging, as it is recalled in the following of the introduction, but which can also be seen as an efficient way to estimate homogenized properties.

X-ray method is commonly used to test the physical properties of materials. By satisfying the Bragg crystal diffraction conditions, the values, the shapes, and the position changes of the diffraction peak are tested to calculate the distance between the crystal surfaces. Then, the strain value at the normal direction of the crystal surface can be solved through the distance between the crystal surfaces mentioned above. Normally, elastic constants measured by X-ray method contain the function of diffraction surface, which can be averaged within the total space to determine the macroscopic elastic properties of the material [7]. By multiplying the relevant X-ray elastic constant and the strain value, the stress in the material can be thus obtained. Until now, for the commonly used X-ray stress analyzer, the principle of X-ray diffraction to test the material stress and to obtain elastic constants [8] is based on the uniform certain stress Reuss model. However, for the uniform certain strain Voigt model, there are still some difficulties in explaining its principle by X-ray diffraction method. Based on the content mentioned above, Y parameter is introduced on the basis of theoretical formulas' derivation to explain both Reuss and Voigt models.

However, for the uniform certain strain Voigt model, there are still some difficulties in explaining its principle by X-ray diffraction method. X-ray method is associated to the material surface, of which X-ray absorption and X-ray magnetic circular dichroism (XMCD) are very powerful tools [9]. Diffraction-based stress analysis depends critically on the use of the correct diffraction elastic constants [10]. By comparing various calculation method to determine homogenized moduli of the polycrystalline material composed of a single crystal, for example, the certain stress of Reuss model [11], the certain strain of Voigt model [12] and Kröner-Voigt model [7] taking into account these two models and interaction between crystals, the Y parameter, in the theoretical calculation to forecast the elastic modulus of polycrystalline material, has a high consistency. Y parameter makes up the methodological disadvantage of X-ray diffraction method (only the stress determined by strain). It can be used to estimate the homogenized properties under various crystal planes, which is the function associated with the crystal plane index.

Y parameters of both cubic and hexagonal polycrystalline materials based on the Voigt model of the certain strain [3, 4] and Y parameters of only hexagonal polycrystalline materials based on the certain stress of the Reuss model [4] have been given. In this work, it is first proposed to complete this part of the theoretical analysis and to obtain the definition of the Y parameter based on the Reuss model [13]. Here, the expression form of the 6×6 matrix (36 parameters) [14] is used instead of the four-rank tensor (81 parameters) in order to simplify the actual computing work and the derivation [15]. Based on the content mentioned above, the concept of Y parameter, as new evaluation criteria, has been proposed on the basis of theoretical formulas' derivation to explain both Reuss and Voigt models.

1.2 Macroscopic mechanical behavior of cubic and hexagonal polycrystals

Polycrystalline material composed of a single crystal, which generally exhibits anisotropic mechanics, with a probability distribution depicting upward almost the same as in three-dimensional space, exhibits an isotropic behavior at macroscopic scale. Using Voigt model, elastic stiffness tensor is obtained by Morris [16] and Sayers [17], using the volume averaging with setting orthogonal constitutive relation of cubic crystal. Based on Voigt and Reuss models, Li and Thompson [18] have deduced elastic stiffness with setting orthogonal constitutive relations of hexagonal crystal for determining the upper and lower bounds. Anderson et al. [19] have verified these bounds and obtained the orthogonal polycrystalline texture coefficient of hexagonal crystal using ultrasonic measurement. Then, the compressibility, shear modulus, bulk modulus, Young's modulus, and Poisson's ratio can be calculated based on the Voigt-Reuss-Hill approximation [20, 21].

Y parameter can be seen as a function of the crystal plane index. Its theoretical derivation is thus relative to the type of crystal. Only cubic and hexagonal structures have been concerned by this concept. Y parameter of cubic and hexagonal structures was first introduced and presented by Zheng et al. [3, 4] in 2009. Elasticity of single crystal and mechanical properties of polycrystalline material has been closely integrated. For that, a preliminary study by ab initio plane-wave pseudopotential density functional theory method is first performed for each of the considered crystalline structures mentioned in Chapter 3; then, this chapter focuses on some cubic, hexagonal, and monoclinic polycrystals to estimate their homogenized elastic moduli. In this part, the methodology of Y parameter is applied to typical cubic, hexagonal, and monoclinic structures, respectively, CaO and MgO, CH, and calcite $CaCO_3$. Elastic moduli of the monoclinic tobermorite and gypsum polycrystals structures are calculated by Reuss-Voigt-Hill estimation [13]. Based on the elastic constants of cubic and hexagonal crystals, the corresponding Y parameter, as the new evaluation criteria, has been proposed. Thus, the credibility of Y parameters for determining elastic moduli of both cubic and hexagonal structures is proved. Elastic moduli of the monoclinic tobermorite structures calculated by Reuss-Voigt-Hill estimation are found to be further investigated for an amorphous structure in Chapter 5.

2. *Y* parameter of cubic structure by Voigt and Reuss models

Voigt and Reuss models are separately based on the assumption of the uniform deformation for upper bound and the uniform stress for lower bound. After that, Hill proposed averaged Voigt and Reuss models with a certain value in engineering application. However, according to the elastic constant definition of Hill model, that is to say $S_{Hill} = \frac{1}{2}\left(S_{\mathrm{Reuss}} + C_{Voigt}^{-1}\right)$ and $C_{Hill} = \frac{1}{2}\left(C_{Voigt} + S_{\mathrm{Reuss}}^{-1}\right)$, it can be seen that the stiffness coefficient C and the compliance coefficient S do not satisfy the relation $C \cdot S = I$. In other words, the elastic constants of Hill model are not self-consistent. That is why the Y parameter has been introduced to redefine the bounds of cubic and hexagonal polycrystals.

For Y-parameter description, macroanalysis of a cubic and hexagonal polycrystalline system is shown in **Figure 1**. At L_3 direction, there exists an orientation relationship between the global coordinate system [S] and the experimental coordinate system [L]. If set the direction cosine of the direction L_3 within system [S] relative to l, m, n, one can get [3]:

$$\begin{aligned} l &= \sin\psi\cos\phi \\ m &= \sin\psi\sin\phi \\ n &= \cos\psi \end{aligned} \tag{1}$$

Therefore, whole strain $\varepsilon_{\phi\psi}$ at the direction L_3 is as [22, 23]:

$$\varepsilon_{\phi\psi} = l^2\varepsilon_{11} + m^2\varepsilon_{22} + n^2\varepsilon_{33} + lm\gamma_{12} + mn\gamma_{23} + nl\gamma_{31} \tag{2}$$

Therefore, whole stress $\sigma_{\phi\psi}$ at the direction L_3 is as [22, 23]:

$$\sigma_{\phi\psi} = l^2\sigma_{11} + m^2\sigma_{22} + n^2\sigma_{33} + 2(lm\tau_{12} + mn\tau_{23} + nl\tau_{31}) \tag{3}$$

For cubic and hexagonal crystals, when the polycrystalline material as a whole is isotropic, Lamé constant λ as a polycrystalline material parameter satisfies the relationship: $\lambda = \frac{2\mu G}{1-2\mu}$, $G = \frac{E}{2(1+\mu)}$; here, E, μ, G are, respectively, Young's modulus,

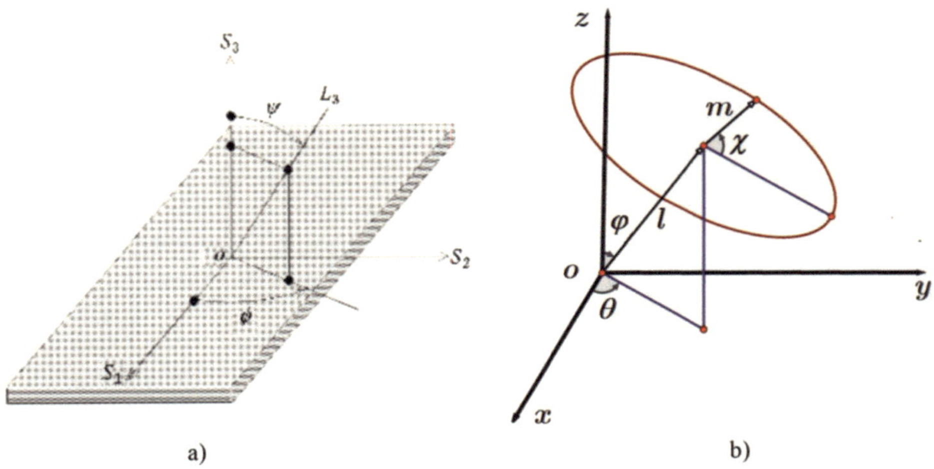

Figure 1.
Macro analysis of a polycrystalline system. (a) Macro analysis of a polycrystalline system and (b) schematic shear crystal plane.

Poisson's ratio, and shear modulus of the polycrystalline materials. When the polycrystalline material as a whole is only under a normal stress, thus $\tau_{12} = \tau_{23} = \tau_{31} = 0$; then, we can see that $\gamma_{12} = \gamma_{23} = \gamma_{31} = 0$. If suppose $\phi = 0^{\circ}$, Eq. (3) can be simplified as [3]:

$$\sigma_{\phi\psi} = l^2\sigma_{11} + m^2\sigma_{22} + n^2\sigma_{33} = \lambda(\varepsilon_{11} + \varepsilon_{22} + \varepsilon_{33}) + 2G\left[(\varepsilon_{11} - \varepsilon_{33})\sin^2\psi + \varepsilon_{33}\right] \tag{4}$$

From **Figure 1**, for Reuss bounds of cubic crystal, when the isotropic material as a whole is only under a uniaxial stress (i.e., $\sigma_{11} \neq 0$, other stress components are zero) and under the conditions of $\phi = 0^{\circ}$, the ratio expression between the whole strain and the uniaxial stress from Eq. (4) will be eventually developed as [3]:

$$\frac{\varepsilon_{\phi\psi}}{\sigma_{11}} = \left(\frac{1+\mu}{E}\right)\sin^2\psi - \left(\frac{\mu}{E}\right) \tag{5}$$

2.1 Reuss bounds of cubic polycrystalline structure

For the Reuss bound of cubic polycrystalline structure, the content is detailed in our previous work [13]. The main crystallographic planes of a cubic crystal are shown in **Figure 2**.

As in **Figure 2**, if the uniaxial tensions are separately applied in the direction of [100], [110], and [111] for a cubic crystal, according to the Ref. [24], Young's moduli by elastic constants C_{11}, C_{12}, and C_{44} can be calculated as follows:

$$E|_{[100]} = \frac{c_{11}^2 + c_{11}c_{12} - 2c_{12}^2}{c_{11} + c_{12}} \tag{6}$$

$$E|_{[110]} = \frac{4\left(c_{11}^2 + c_{11}c_{12} - 2c_{12}^2\right)c_{44}}{c_{11}^2 + c_{11}c_{12} + 2c_{11}c_{44} - 2c_{12}^2} \tag{7}$$

$$E|_{[111]} = \frac{3(c_{11} + 2c_{12})c_{44}}{c_{11} + 2c_{12} + c_{44}} \tag{8}$$

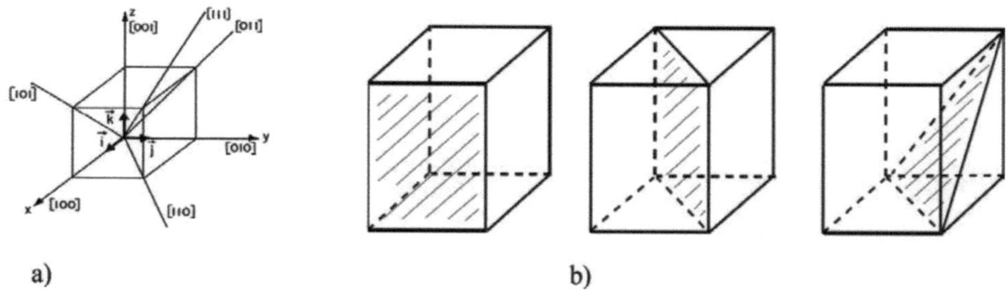

Figure 2.
Some crystallographic planes of a cubic crystal. (a) The directions in a cubic lattice and (b) the {100}, {110}, and {111} planes.

For cubic crystal, Y elastic constants based on Reuss model which has contained crystal surface normal ($n = (u, v, w)$) are as:

$$G_Y(u,v,w) = \frac{1}{2(s_{11}-s_{12}) - 3[2(s_{11}-s_{12}) - s_{44}]\Gamma} \tag{9}$$

$$E_Y(u,v,w) = \frac{1}{s_{11} - [2(s_{11}-s_{12}) - s_{44}]\Gamma} \tag{10}$$

$$\mu_Y(u,v,w) = -\frac{2s_{12} + (2(s_{11}-s_{12}) - s_{44})\Gamma}{2\{s_{11} - [2(s_{11}-s_{12}) - s_{44}]\Gamma\}} \tag{11}$$

where $\overline{\Gamma} = u^2v^2 + v^2w^2 + w^2u^2$. From Eqs. (9)–(11), elastic moduli of cubic structures by Y parameters can be calculated, and the relationship of G-3Γ, E-3Γ, and μ-3Γ can be calculated and the limitation of elastic moduli can be determined by Reuss and Voigt models.

2.2 Voigt bounds of cubic polycrystalline structure

As is shown in **Figure 2**, supposing that the direction cosines are L_3 (u, v, w) relative to the cubic crystal coordinates $[K]$, the relationship between the macroscopic stress σ^{L3} in L_3 direction and the stress component $(\sigma_x, \sigma_y, \sigma_z, \tau_{yz}, \tau_{zx}, \tau_{xy})$ of the coordinate system $[K]$ is as [22, 23]:

$$\sigma^{L_3} = u^2\sigma_x + v^2\sigma_y + w^2\sigma_z + 2(vw\tau_{yz} + wu\tau_{zx} + uv\tau_{xy}) \tag{12}$$

With respect to the coordinate system $[K]$, the relationship between stress and strain of single cubic crystal structure is as:

$$\begin{bmatrix} \sigma_x \\ \sigma_y \\ \sigma_z \\ \tau_{yz} \\ \tau_{zx} \\ \tau_{xy} \end{bmatrix} = \begin{bmatrix} c_{11} & c_{12} & c_{12} & 0 & 0 & 0 \\ c_{12} & c_{11} & c_{12} & 0 & 0 & 0 \\ c_{12} & c_{12} & c_{11} & 0 & 0 & 0 \\ 0 & 0 & 0 & c_{44} & 0 & 0 \\ 0 & 0 & 0 & 0 & c_{44} & 0 \\ 0 & 0 & 0 & 0 & 0 & c_{44} \end{bmatrix} \begin{bmatrix} \varepsilon_x \\ \varepsilon_y \\ \varepsilon_z \\ \gamma_{yz} \\ \gamma_{zx} \\ \gamma_{xy} \end{bmatrix} \tag{13}$$

Submitting Eq. (13) into Eq. (12), one can get:

$$\sigma^{L_3} = c_{11}\left(u^2\varepsilon_x + v^2\varepsilon_y + w^2\varepsilon_z\right) + c_{12}\left[\left(v^2 + w^2\right)\varepsilon_x + \left(u^2 + w^2\right)\varepsilon_y + \left(u^2 + v^2\right)\varepsilon_z\right] + 2c_{44}\left(vw\gamma_{yz} + uw\gamma_{zx} + uv\gamma_{xy}\right) \tag{14}$$

With respect to the coordinate system $[K]$, assuming the three direction cosines of coordinate axes [S] (respectively S_1, S_2, S_3) are S_1 (l_1, m_1, n_1), S_2 (l_2, m_2, n_2), S_3 (l_3, m_3, n_3), and considering relationship between strain components in coordinates system [S] and strain component of coordinates system $[K]$, we can get relation $\gamma_{12} = \gamma_{23} = \gamma_{31} = 0$ when the polycrystalline material is limited only by the principal stress $(\tau_{12} = \tau_{23} = \tau_{31} = 0)$. So the following formula of strain holds [22]:

$$\begin{aligned}
\varepsilon_x &= l_1^2\varepsilon_{11} + l_2^2\varepsilon_{22} + l_3^2\varepsilon_{33} \\
\varepsilon_y &= m_1^2\varepsilon_{11} + m_2^2\varepsilon_{22} + m_3^2\varepsilon_{33} \\
\varepsilon_z &= n_1^2\varepsilon_{11} + n_2^2\varepsilon_{22} + n_3^2\varepsilon_{33} \\
\gamma_{yz} &= 2(m_1n_1\varepsilon_{11} + m_2n_2\varepsilon_{22} + m_3n_3\varepsilon_{33}) \\
\gamma_{zx} &= 2(n_1l_1\varepsilon_{11} + n_2l_2\varepsilon_{22} + n_3l_3\varepsilon_{33}) \\
\gamma_{xy} &= 2(l_1m_1\varepsilon_{11} + l_2m_2\varepsilon_{22} + l_3m_3\varepsilon_{33})
\end{aligned} \tag{15}$$

Substituting Eq. (15) into Eq. (14), one can get:

$$\begin{aligned}
\sigma^{L_3} = {} & c_{11}\left[u^2\left(l_1^2\varepsilon_{11} + l_2^2\varepsilon_{22} + l_3^2\varepsilon_{33}\right) + v^2\left(m_1^2\varepsilon_{11} + m_2^2\varepsilon_{22} + m_3^2\varepsilon_{33}\right) + w^2\left(n_1^2\varepsilon_{11} + n_2^2\varepsilon_{22} + n_3^2\varepsilon_{33}\right)\right] \\
& + c_{12}\left[\left(v^2 + w^2\right)\left(l_1^2\varepsilon_{11} + l_2^2\varepsilon_{22} + l_3^2\varepsilon_{33}\right) + \left(u^2 + w^2\right)\left(m_1^2\varepsilon_{11} + m_2^2\varepsilon_{22} + m_3^2\varepsilon_{33}\right)\right. \\
& \left. + \left(u^2 + v^2\right)\left(n_1^2\varepsilon_{11} + n_2^2\varepsilon_{22} + n_3^2\varepsilon_{33}\right)\right] \\
& + 4c_{44}\left[vw(m_1n_1\varepsilon_{11} + m_2n_2\varepsilon_{22} + m_3n_3\varepsilon_{33}) + uw(n_1l_1\varepsilon_{11} + n_2l_2\varepsilon_{22} + n_3l_3\varepsilon_{33})\right. \\
& \left. + uv(l_1m_1\varepsilon_{11} + l_2m_2\varepsilon_{22} + l_3m_3\varepsilon_{33})\right]
\end{aligned} \tag{16}$$

In practice, for cubic polycrystalline structure, the single crystal has a haphazard distribution, and the probability distribution of the crystals depicting up almost the same in the three-dimensional space. So, we can average all directions in three-dimensional space, and then obtain the mean stress $\overline{\sigma^{L_3}}$ so as to determine its macroscopic stress. Here, if $\phi = 0^{\circ}$, we can obtain the mean stress $\overline{\sigma_{Y^{L_3}}}$ by averaging the orientation in the crystal plane:

$$\overline{\sigma_Y^{L_3}} = \int_0^{2\pi} \sigma^{L_3} d\theta \Big/ \int_0^{2\pi} d\theta \tag{17}$$

Here, we use the certain strain of Voigt model, where the material is only under the uniaxial strain [12]. For polycrystalline, the stress can be obtained by averaging integral within a crystal surface according to the formula (17):

$$\begin{aligned}
\overline{\sigma_Y^{L_3}} = {} & [c_{11} - c_{12} - 3(c_{11} - c_{12} - 2c_{44})\Gamma](\varepsilon_{11} - \varepsilon_{33})\sin^2\psi + [c_{11} - c_{12} - 3(c_{11} - c_{12} - 2c_{44})\Gamma]\varepsilon_{33} \\
& + (\varepsilon_{11} + \varepsilon_{22} + \varepsilon_{33}) \cdot [c_{12} + (c_{11} - c_{12} - 2c_{44})\Gamma]
\end{aligned} \tag{18}$$

where $\Gamma = u^2v^2 + v^2w^2 + u^2w^2$, (u, v, w) is the cosine normals within planes of cubic crystal. With comparison of Eq. (18) and Eq. (16), we can obtain Y elastic constants containing crystal surface normal ($n = (u, v, w)$) as [3]:

$$G_Y(u,v,w) = \frac{c_{11}-c_{12}}{2} - \frac{3(c_{11}-c_{12}-2c_{44})\Gamma}{2} \tag{19}$$

$$\lambda_Y(u,v,w) = c_{12} + (c_{11}-c_{12}-2c_{44})\Gamma \tag{20}$$

As $\lambda = \frac{2\mu G}{1-2\mu}$, $G = \frac{E}{2(1+\mu)}$, formulas can be developed as:

$$\left(\frac{1+\mu}{E}\right)_Y = \frac{1}{c_{11}-c_{12}-(c_{11}-c_{12}-2c_{44})\cdot 3\Gamma} \tag{21}$$

$$-\left(\frac{\mu}{E}\right)_Y = -\frac{c_{12}+(c_{11}-c_{12}-2c_{44})\Gamma}{(c_{11}+2c_{12})[c_{11}-c_{12}-3(c_{11}-c_{12}-2c_{44})\Gamma]} \tag{22}$$

$$E_Y(u,v,w) = \frac{(c_{11}+2c_{12})[c_{11}-c_{12}-3(c_{11}-c_{12}-2c_{44})\Gamma]}{c_{11}+c_{12}-(c_{11}-c_{12}-2c_{44})\Gamma} \tag{23}$$

$$\mu_Y(u,v,w) = \frac{c_{12}+(c_{11}-c_{12}-2c_{44})\Gamma}{c_{11}+c_{12}-(c_{11}-c_{12}-2c_{44})\Gamma} \tag{24}$$

Mechanical properties of polycrystalline can be obtained by Y elastic constant with averaging in the (u, v, w) entire 3-D space orientation, which are all elastic parameters having nothing to do with the crystal surface index. Assuming $\overline{\Gamma} = 1/5$ [25], the mechanical properties of polycrystalline, such as Young's modulus E, Poisson's ratio μ, and the shear modulus G, are as [3]:

$$G_M = \frac{c_{11}-c_{12}+3c_{44}}{5}, \lambda_M = \frac{c_{11}+4c_{12}-2c_{44}}{5} \tag{25}$$

$$\left(\frac{1+\mu}{E}\right)_M = \frac{5}{2(c_{11}-c_{12}+3c_{44})} \tag{26}$$

$$-\left(\frac{\mu}{E}\right)_M = -\frac{c_{11}+4c_{12}-2c_{44}}{2(c_{11}+2c_{12})(c_{11}-c_{12}+3c_{44})} \tag{27}$$

$$E_M = \frac{(c_{11}+2c_{12})(c_{11}-c_{12}+3c_{44})}{2c_{11}+3c_{12}+c_{44}} \tag{28}$$

$$\mu_M = \frac{c_{11}+4c_{12}-2c_{44}}{2(2c_{11}+3c_{12}+c_{44})} \tag{29}$$

Eqs. (19)–(24) are the theoretical formula of mechanical elastic constants and Y parameters of cubic polycrystal structures based on a certain strain of Voigt model.

For cubic polycrystalline model, at first, we set a face normal as the basis axis on surface of a polycrystalline, and then average the normals in all 360° orientation within the polycrystalline surface. After obtaining Y elastic constants, we choose the average orientation of crystal surface normals on the entire three-dimensional space.

3. *Y* parameter of hexagonal structure by Voigt and Reuss models

Similarly for hexagonal crystal from **Figure 1**, for Voigt bounds, when the isotropic material as a whole is only under a uniaxial stress (i.e., $\sigma_{11} \neq 0$, other stress components are zero) and under the conditions of $\phi = 0^\circ$, the ratio expression between the whole strain and the uniaxial stress from Eq. (4) will be eventually developed as [4]:

$$\sigma_{\phi\psi} = \lambda(\varepsilon_{11}+\varepsilon_{22}+\varepsilon_{33}) + 2G\left[(\varepsilon_{11}-\varepsilon_{33})\sin^2\psi + \varepsilon_{33}\right] \tag{30}$$

3.1 Reuss bounds of hexagonal polycrystal structure

As in **Figure 1**, supposing that the direction cosines are L_3 (u, v, w) relative to the hexagonal crystal coordinates $[K]$, the relationship between the macroscopic strain ε^{L_3} (equivalent to $\varepsilon_{\phi\psi}$) in L_3 direction and the strain component $(\varepsilon_x, \varepsilon_y, \varepsilon_z, \gamma_{xy}, \gamma_{yz}, \gamma_{zx})$ of the coordinate system $[K]$ is as [22, 23]:

$$\varepsilon^{L_3} = u^2\varepsilon_x + v^2\varepsilon_y + w^2\varepsilon_z + uv\gamma_{xy} + vw\gamma_{yz} + wu\gamma_{zx} \tag{31}$$

With respect to the coordinate system $[K]$, the relationship between strain and stress of a single hexagonal crystal structure is as [23]:

$$\begin{bmatrix} \varepsilon_x \\ \varepsilon_y \\ \varepsilon_z \\ \gamma_{yz} \\ \gamma_{zx} \\ \gamma_{xy} \end{bmatrix} = \begin{bmatrix} s_{11} & s_{12} & s_{13} & 0 & 0 & 0 \\ s_{12} & s_{11} & s_{13} & 0 & 0 & 0 \\ s_{13} & s_{13} & s_{33} & 0 & 0 & 0 \\ 0 & 0 & 0 & s_{44} & 0 & 0 \\ 0 & 0 & 0 & 0 & s_{44} & 0 \\ 0 & 0 & 0 & 0 & 0 & 2(s_{11} - s_{12}) \end{bmatrix} \begin{bmatrix} \sigma_x \\ \sigma_y \\ \sigma_z \\ \tau_{yz} \\ \tau_{zx} \\ \tau_{xy} \end{bmatrix} \tag{32}$$

Substituting Eq. (32) into Eq. (31), one can get:

$$\begin{aligned} \varepsilon^{L_3} &= s_{11}\left(u^2\sigma_x + v^2\sigma_y + 2uv\tau_{xy}\right) + s_{12}\left(u^2\sigma_y + v^2\sigma_x - 2uv\tau_{xy}\right) \\ &+ s_{13}\left[\left(u^2+v^2\right)\sigma_z + w^2\left(\sigma_x +^x \sigma_y\right)\right] + s_{33}w^2\sigma_z + s_{44}w\left(v\tau_{yz} + u\tau_{zx}\right) \end{aligned} \tag{33}$$

With respect to the coordinate system $[K]$, assuming the three direction cosines of coordinate axes $[S]$ (respectively S_1, S_2, S_3) are S_1 (l_1, m_1, n_1), S_2 (l_2, m_2, n_2), S_3 (l_3, m_3, n_3), the relationship between the stress components in the coordinate system $[K]$ and the stress component of the coordinate system $[S]$ is as follows [26]:

$$\sigma_x = \sum l_i l_j \sigma_{ij} \tag{34}$$

$$\sigma_y = \sum m_i m_j \sigma_{ij} \tag{35}$$

$$\sigma_z = \sum n_i n_j \sigma_{ij} \tag{36}$$

$$\tau_{xy} = \sum l_i m_j \sigma_{ij} \tag{37}$$

$$\tau_{yz} = \sum m_i n_j \sigma_{ij} \tag{38}$$

$$\tau_{zx} = \sum n_i l_j \sigma_{ij} \tag{39}$$

When the material is only under the uniaxial stress $\sigma_{11} \neq 0$ and other stress components are zero, under these conditions, substituting Eqs. (34)–(39) into Eq. (54), one can get:

$$\begin{aligned} \varepsilon^{L_3} = &\left\{s_{11}\left(u^2 l_1^2 + v^2 m_1^2 + 2uvl_1 m_1\right) + s_{12}\left(u^2 m_1^2 + v^2 l_1^2 - 2uvl_1 m_1\right) + \right. \\ &\left. s_{13}\left[\left(u^2+v^2\right)n_1^2 + w^2\left(l_1^2 + m_1^2\right)\right] + s_{33}w^2 n_1^2 + s_{44}w\left(vm_1 n_1 + ul_1 n_1\right)\right\}\sigma_{11} \end{aligned} \tag{40}$$

In practice, for hexagonal polycrystalline structure, the single crystal has a haphazard distribution, and the probability distribution of the crystals depicting up almost the same in the three-dimensional space. So, we can average all directions in three-dimensional space, and then obtain the mean strain $\overline{\varepsilon^{L_3}}$ so as to determine its macroscopic strain. Here, if $\phi = 0^\circ$, we can obtain the mean strain $\overline{\varepsilon_{\gamma L_3}}$ by averaging the orientation in the crystal plane:

$$\overline{\varepsilon_Y^{L_3}} = \int_0^{2\pi} \varepsilon^{L_3} d\theta \Big/ \int_0^{2\pi} d\theta \tag{41}$$

Here, (u, v, w) are normal direction cosines of hexagonal crystal surface, so we take Reuss model using a certain stress. For polycrystalline, the stress can be obtained by averaging integral within a crystal surface according to the formula (41):

$$\frac{\overline{\varepsilon_Y^{L_3}}}{\sigma_{11}} = \left[\frac{1}{2}(2s_{11} - s_{12} - s_{13}) - \frac{w^2}{2}(5s_{11} - s_{12} - 5s_{13} + s_{33} - 3s_{44}) + \frac{3w^4}{2}(s_{11} - 2s_{13} + s_{33} - s_{44})\right]$$
$$\times \sin^2\psi + \frac{1}{2}(s_{12} + s_{13}) + \frac{w^2}{2}(s_{11} - s_{12} - s_{13} + s_{33} - s_{44}) - \frac{w^4}{2}(s_{11} - 2s_{13} + s_{33} - s_{44}) \tag{42}$$

With comparison of Eq. (42) and Eq. (33), we can obtain Y elastic constants containing crystal surface normal (n = (u, v, w)) as:

$$\left(\frac{1+\mu}{E}\right)_Y = \frac{1}{2}(2s_{11} - s_{12} - s_{13}) - \frac{w^2}{2}(5s_{11} - s_{12} - 5s_{13} + s_{33} - 3s_{44}) + \frac{3w^4}{2}(s_{11} - 2s_{13} + s_{33} - s_{44}) \tag{43}$$

$$-\left(\frac{\mu}{E}\right)_Y = \frac{1}{2}(s_{12} + s_{13}) + \frac{w^2}{2}(s_{11} - s_{12} - s_{13} + s_{33} - s_{44}) - \frac{w^4}{2}(s_{11} - 2s_{13} + s_{33} - s_{44}) \tag{44}$$

As, according to Eq. (43) and Eq. (44), we can get

$$E_Y(u, v, w) = \frac{1}{s_{11} - [2(s_{11} - s_{13}) - s_{44}]w^2 + (s_{11} - 2s_{13} + s_{33} - s_{44})w^4} \tag{45}$$

$$\mu_Y(u, v, w) = -\frac{s_{12} + s_{13} + (s_{11} - s_{12} - s_{13} + s_{33} - s_{44})w^2 - (s_{11} - 2s_{13} + s_{33} - s_{44})w^4}{2\{s_{11} - [2(s_{11} - s_{13}) - s_{44}]w^2 + (s_{11} - 2s_{13} + s_{33} - s_{44})w^4\}} \tag{46}$$

$$G_Y(u, v, w) = \frac{1}{(2s_{11} - s_{12} - s_{13}) - (5s_{11} - s_{12} - 5s_{13} + s_{33} - 3s_{44})w^2 + 3(s_{11} - 2s_{13} + s_{33} - s_{44})w^4} \tag{47}$$

Mechanical properties of polycrystalline can be obtained by Y elastic constant with averaging in the (u, v, w) entire 3-D space orientation, which are all elastic parameters having nothing to do with the crystal surface index. Assuming $\overline{w^2} = \frac{1}{3}$ and $\overline{w^4} = \frac{1}{5}$ [25], the mechanical properties of polycrystalline, such as Young's modulus E, Poisson's ratio μ, and the shear modulus G, are as [4]:

$$\left(\frac{1+\mu}{E}\right)_M = \frac{7s_{11} - 5s_{12} - 4s_{13} + 2s_{33} + 3s_{44}}{15} \tag{48}$$

$$-\left(\frac{\mu}{E}\right)_M = \frac{s_{11} + 5s_{12} + 8s_{13} + s_{33} - s_{44}}{15} \tag{49}$$

$$E = \frac{15}{2(4s_{11} + 2s_{13} + s_{44}) + 3s_{33}} \tag{50}$$

$$\mu = \frac{s_{11} + 5s_{12} + 8s_{13} + s_{33} - s_{44}}{2(4s_{11} + 2s_{13} + s_{44}) + 3s_{33}} \tag{51}$$

$$G = \frac{15}{2(7s_{11} - 5s_{12} - 4s_{13} + 2s_{33} + 3s_{44})} \tag{52}$$

Based on the Reuss bounds and elastic constants of a single crystal, the theoretical formulas (48)–(52) of mechanical elastic moduli of hexagonal polycrystalline are finally determined.

3.2 Voigt bounds of hexagonal polycrystal structure

With respect to the coordinate system [K], the relationship between strain and stress of a single hexagonal crystal is as in Eq. (54). By Eq. (54) and Ref. [3], one can get:

$$\begin{aligned} \sigma^{L_3} &= c_{11}\left(u^2\varepsilon_x + v^2\varepsilon_y + uv\gamma_{xy}\right) + c_{12}\left(u^2\varepsilon_y + v^2\varepsilon_x - uv\gamma_{xy}\right) \\ &+ c_{13}\left[(u^2 + v^2)\varepsilon_z + w^2\left(\varepsilon_x +^x \varepsilon_y\right)\right] + c_{33}w^2\varepsilon_z + 2c_{44}w\left(v\gamma_{yz} + u\gamma_{zx}\right) \end{aligned} \tag{53}$$

Similar to Ref. [3], according to Voigt model [12], we can average oriented polycrystalline material within a crystal surface and obtain the mean stress:

$$\begin{aligned} \overline{\sigma_Y^{L_3}} &= \frac{1}{2}\left[2c_{11} - c_{12} - c_{13} - w^2(5c_{11} - c_{12} - 5c_{13} + c_{33} - 12c_{44}) + 3w^4(c_{11} - 2c_{13} + c_{33} - 4c_{44})\right] \\ &\cdot (\varepsilon_{11} - \varepsilon_{33}) \sin^2\psi \\ &+ \frac{1}{2}\left[2c_{11} - c_{12} - c_{13} - w^2(5c_{11} - c_{12} - 5c_{13} + c_{33} - 12c_{44}) + 3w^4(c_{11} - 2c_{13} + c_{33} - 4c_{44})\right] \\ &\cdot \varepsilon_{33} + \frac{1}{2}(\varepsilon_{11} + \varepsilon_{22} + \varepsilon_{33}) \\ &\cdot \left[c_{12} + c_{13} + w^2(c_{11} - c_{12} - c_{13} + c_{33} - 4c_{44}) - w^4(c_{11} - 2c_{13} + c_{33} - 4c_{44})\right] \end{aligned} \tag{54}$$

According to Eq. (53) and Ref. [3], we can obtain Y elastic constants containing crystal surface normal ($n = (u, v, w)$) as:

$$G_Y(u,v,w) = \frac{1}{4}\left[2c_{11} - c_{12} - c_{13} - w^2(5c_{11} - c_{12} - 5c_{13} + c_{33} - 12c_{44}) + 3w^4(c_{11} - 2c_{13} + c_{33} - 4c_{44})\right] \tag{55}$$

$$\lambda_Y(u,v,w) = \frac{1}{4}\left[c_{12} + c_{13} + w^2(c_{11} - c_{12} - c_{13} + c_{33} - 4c_{44}) - w^4(c_{11} - 2c_{13} + c_{33} - 4c_{44})\right] \tag{56}$$

As $G = \frac{E}{2(1+\mu)}$, $\lambda = \frac{2\mu G}{1-2\mu}$, formulas can be developed as follows [4]:

$$\left(\frac{1+\mu}{E}\right)_Y = \frac{2}{2c_{11} - c_{12} - c_{13} - w^2(5c_{11} - c_{12} - 5c_{13} + c_{33} - 12c_{44}) + 3w^4(c_{11} - 2c_{13} + c_{33} - 4c_{44})} \tag{57}$$

$$-\left(\frac{\mu}{E}\right)_Y = -\frac{c_{12} + c_{13} + w^2(c_{11} - c_{12} - c_{13} + c_{33} - 4c_{44}) - w^4(c_{11} - 2c_{13} + c_{33} - 4c_{44})}{\left[c_{11} + c_{12} + c_{13} - w^2(c_{11} + c_{12} - c_{13} - c_{33})\right] \cdot \left[\begin{matrix} 2c_{11} - c_{12} - c_{13} - w^2(5c_{11} - c_{12} - 5c_{13} + c_{33} - 12c_{44}) \\ +3w^4(c_{11} - 2c_{13} + c_{33} - 4c_{44}) \end{matrix}\right]} \tag{58}$$

$$\begin{aligned} E_Y(u,v,w) &= \left[c_{11} + c_{12} + c_{13} - w^2(c_{11} + c_{12} - c_{13} - c_{33})\right] \cdot \\ &\frac{\left[2c_{11} - c_{12} - c_{13} - w^2(5c_{11} - c_{12} - 5c_{13} + c_{33} - 12c_{44}) + 3w^4(c_{11} - 2c_{13} + c_{33} - 4c_{44})\right]}{2c_{11} + c_{12} + c_{13} - w^2(3c_{11} + c_{12} - 3c_{13} - c_{33} - 4c_{44}) + w^4(c_{11} - 2c_{13} + c_{33} - 4c_{44})} \end{aligned} \tag{59}$$

$$\mu_Y(u,v,w) = \frac{c_{12}+c_{13}+w^2(c_{11}-c_{12}-c_{13}+c_{33}-4c_{44})-w^4(c_{11}-2c_{13}+c_{33}-4c_{44})}{2c_{11}+c_{12}+c_{13}-w^2(3c_{11}+c_{12}-3c_{13}-c_{33}-4c_{44})+w^4(c_{11}-2c_{13}+c_{33}-4c_{44})} \tag{60}$$

Mechanical properties of polycrystalline can be obtained by Y elastic constant with averaging in the (u, v, w) entire *3-D* space orientation, which are all elastic parameters having nothing to do with the crystal surface index. Assuming $\overline{w^2} = \frac{1}{3}$ and $\overline{w^4} = \frac{1}{5}$ [25], the mechanical properties of polycrystalline, such as Young's modulus E, Poisson's ratio μ, and shear modulus G, are as [4]:

$$G_M = \frac{7c_{11}-5c_{12}-4c_{13}+2c_{33}+12c_{44}}{30} \tag{61}$$

$$\lambda_M = \frac{c_{11}+5c_{12}+8c_{13}+c_{33}-4c_{44}}{15} \tag{62}$$

$$\left(\frac{1+\mu}{E}\right)_M = \frac{15}{7c_{11}-5c_{12}-4c_{13}+2c_{33}+12c_{44}} \tag{63}$$

$$-\left(\frac{\mu}{E}\right)_M = \frac{3(c_{11}+5c_{12}+8c_{13}+c_{33}-4c_{44})}{[2(c_{11}+c_{12}+2c_{13})+c_{33}](7c_{11}-5c_{12}-4c_{13}+2c_{33}+12c_{44})} \tag{64}$$

$$E = \frac{[2(c_{11}+c_{12}+2c_{13})+c_{33}](7c_{11}-5c_{12}-4c_{13}+2c_{33}+12c_{44})}{3[9c_{11}+5c_{12}+4(3c_{13}+c_{33}+c_{44})]} \tag{65}$$

$$\mu = \frac{c_{11}+5c_{12}+8c_{13}+c_{33}-4c_{44}}{9c_{11}+5c_{12}+4(3c_{13}+c_{33}+c_{44})} \tag{66}$$

Based on the Voigt bounds, elastic constants of a single crystal and the theoretical formulas (61)–(66), mechanical elastic moduli of polycrystalline hexagonal structure are finally determined.

For the relative formulas of cubic and hexagonal material, the elastic moduli (Young's modulus E_φ and shear modulus G_φ) can be determined along any orientation, from the elastic constants or compliance constants [13].

For monoclinic structures, the relationship between elastic compliance and Young's modulus E_Φ in either direction of crystal c-axis with the angle Φ can be described as [1]:

$$\begin{aligned}\frac{1}{E_\phi} &= S_{11}l_1^4 + 2S_{12}l_1^2l_2^2 + 2S_{13}l_1^2l_3^2 + 2S_{15}l_1^3l_3 + S_{22}l_2^4 + 2S_{23}l_2^2l_3^2 + 2S_{25}l_1l_2^2l_3 \\ &+ S_{33}l_3^4 + 2S_{35}l_1l_3^3 + S_{44}l_2^2l_3^2 + 2S_{46}l_1l_2^2l_3 + S_{55}l_1^2l_3^2 + S_{66}l_1^2l_2^2\end{aligned} \tag{67}$$

where $\begin{pmatrix} l_1 \\ l_2 \\ l_3 \end{pmatrix} = \begin{pmatrix} \sin\theta\cos\phi \\ \sin\theta\sin\phi \\ \cos\theta \end{pmatrix}$, S_{ij} is the single crystal compliance tensor in collapsed matrix notation and l_1, l_2, and l_3 are the direction cosines between the direction normal to the *hkl* plane and the lattice vectors (a, b, and c). So, for a given *hkl* plane, the elastic modulus E_{hkl} in a normal direction to *hkl* plane in polycrystalline can be determined from Eq. (67).

Meanwhile, for cubic and hexagonal material, the interchangeable relationship between elastic stiffness constants C_{ij} and the elastic compliance constants S_{ij} is detailed in Ref. [13].

4. Elastic moduli of cubic polycrystals structures at nanoscale

The homogenized moduli of typical cubic polycrystals can be calculated by Reuss-Voigt-Hill estimation [13].

4.1 Elastic moduli of typical cubic CaO structure

Elastic constants of CaO single crystal by GGA are as: c_{11} = 200.879 GPa, c_{12} = 56.902 GPa, and c_{44} = 75.600 GPa. By LDA, the C_{ij} are as: c_{11} = 289.497 GPa, c_{12} = 71.432 GPa, and c_{44} = 76.652 GPa. The elastic moduli of CaO polycrystalline based on Y parameters were calculated, shown in **Figure 3**.

Mechanical moduli of CaO polycrystalline by Reuss-Voigt-Hill method are shown in **Table 1**.

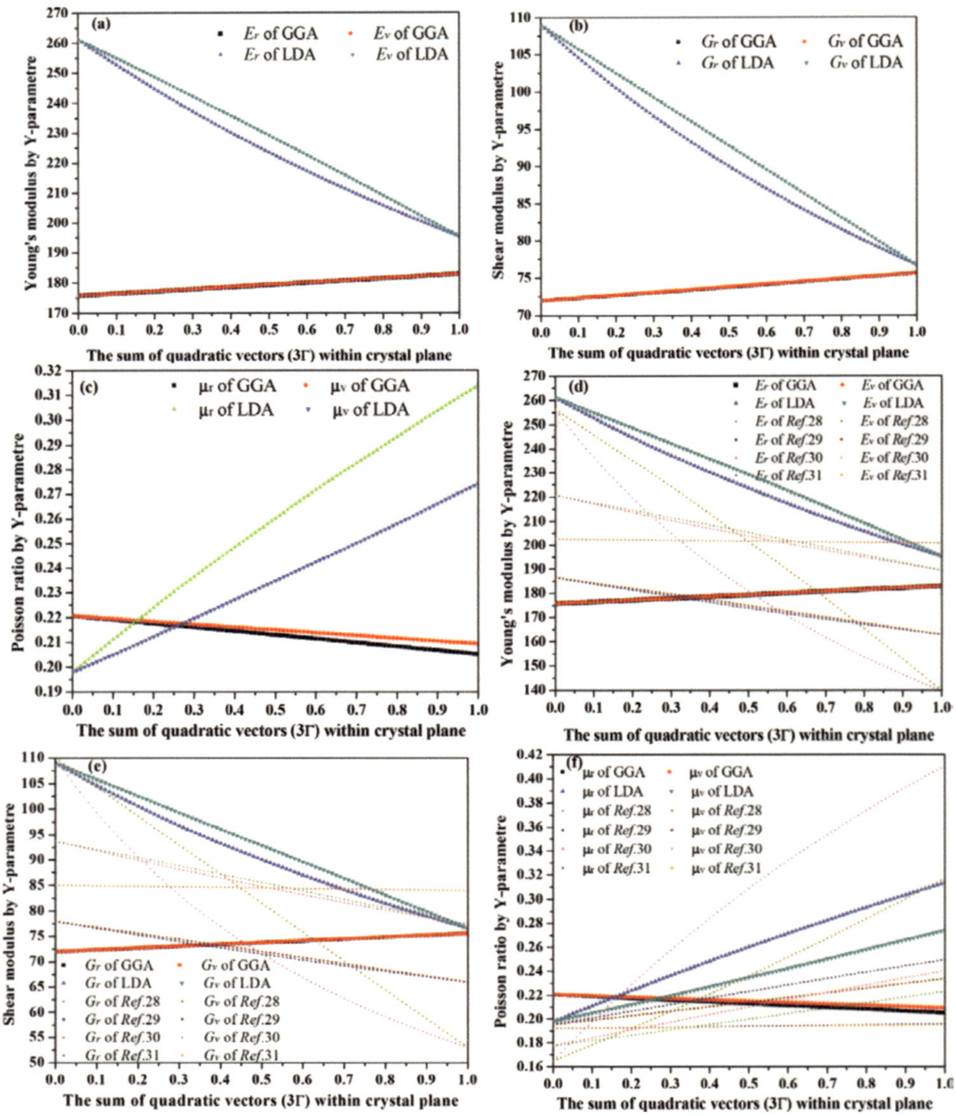

Figure 3.
Poisson ratio, Young's, and shear moduli of CaO by Y parameter.

P	B_v	B_r	G_v	G_r	*B*	*G*	*E*	μ
Ref. [27]	127.3333	127.3333	75.8000	66.8578	127.3333	71.3289	180.3170	0.2640
Ref. [28]	102	102	70.8000	70.3279	102	70.5639	172.0231	0.2189
Ref. [29]	114.0667	114.0667	83.9200	83.1886	114.0667	83.5543	201.4703	0.2056
Ref. [30]	109.6667	109.6667	84.4000	84.3972	109.6667	84.3986	201.5038	0.1938
0 (GGA)	104.894	104.894	74.155	74.113	104.894	74.134	179.9978	0.2140
0 (LDA)	144.120	144.120	89.604	86.985	144.120	88.295	219.9640	0.2456
0.1 (GGA)	105.011	105.011	74.334	74.300	105.011	74.317	176.617	0.2197
0.2 (GGA)	105.503	105.503	74.494	74.467	105.503	74.481	177.512	0.2196
0.3 (GGA)	105.616	105.616	74.629	74.608	105.616	74.619	178.159	0.2189
0.4 (GGA)	106.409	106.409	74.839	74.825	106.409	74.832	179.377	0.2190
0.5 (GGA)	107.046	107.046	75.016	75.006	107.046	75.011	180.357	0.2192
0.6 (GGA)	106.995	106.995	75.163	75.157	106.995	75.160	181.067	0.2180
0.7 (GGA)	107.443	107.443	75.319	75.315	107.443	75.317	181.873	0.2179
0.8 (GGA)	107.908	107.908	75.558	75.557	107.908	75.558	183.113	0.2172
0.9 (GGA)	108.600	108.600	75.730	75.730	108.600	75.730	184.108	0.2175
1.0 (GGA)	108.650	108.650	75.848	75.848	108.650	75.848	184.664	0.2167

Table 1.
Mechanical moduli of CaO polycrystalline by different methods.

From **Figure 3(a)–(c)**, for CaO polycrystals, Y parameters of Reuss and Voigt models are very close to each other based on the elastic constants by GGA method. Besides, the Poisson ratio, Young's, and shear moduli by GGA method have a very good agreement with other Refs. [27–30]. By Y parameter, shear modulus is between 71.99 and 75.61 GPa, while Young's modulus is between 175.78 and 182.86 GPa under overall crystal plane orientation.

From **Table 1**, the elastic moduli *B* and *G* of CaO can be obtained based on the dependences of the elastic constants C_{ij}. Mechanical moduli of CaO polycrystalline by different methods are shown in **Table 1**. Elastic moduli at 0 GPa are verified and averaged as: B = 104.894 GPa, G = 74.134 GPa, and μ = 0.2209 in **Table 1**. Young's modulus is about 175.758 GPa by Reuss-Voigt-Hill estimation.

According to Eqs. (26)–(28), Young's moduli of CaO at 0 GPa in directions of *[100]*, *[110]*, and *[111]* are determined as: $E_{[100]}$ = 175.758 GPa, $E_{[110]}$ = 181.037 GPa, and $E_{[111]}$ = 182.868 GPa.

4.2 Elastic moduli of typical cubic MgO structure

Y parameters calculated by the polycrystalline bounds of MgO curves are shown in **Figure 4**.

As can be seen from **Figure 4(a)–(c)**, MgO polycrystals, Y parameters of Reuss and Voigt models are very close to each other based on the elastic constants by GGA method. Besides, the Poisson ratio, Young's, and shear moduli by GGA method have a very good agreement with other Refs. [28, 30–32]. Moreover, shear modulus of MgO by Reuss and Voigt bounds of Y parameter is between 118.78 and 140.96 GPa under overall crystal plane orientation. Moreover, Young's modulus of MgO is between 293.62 and 337.21 GPa.

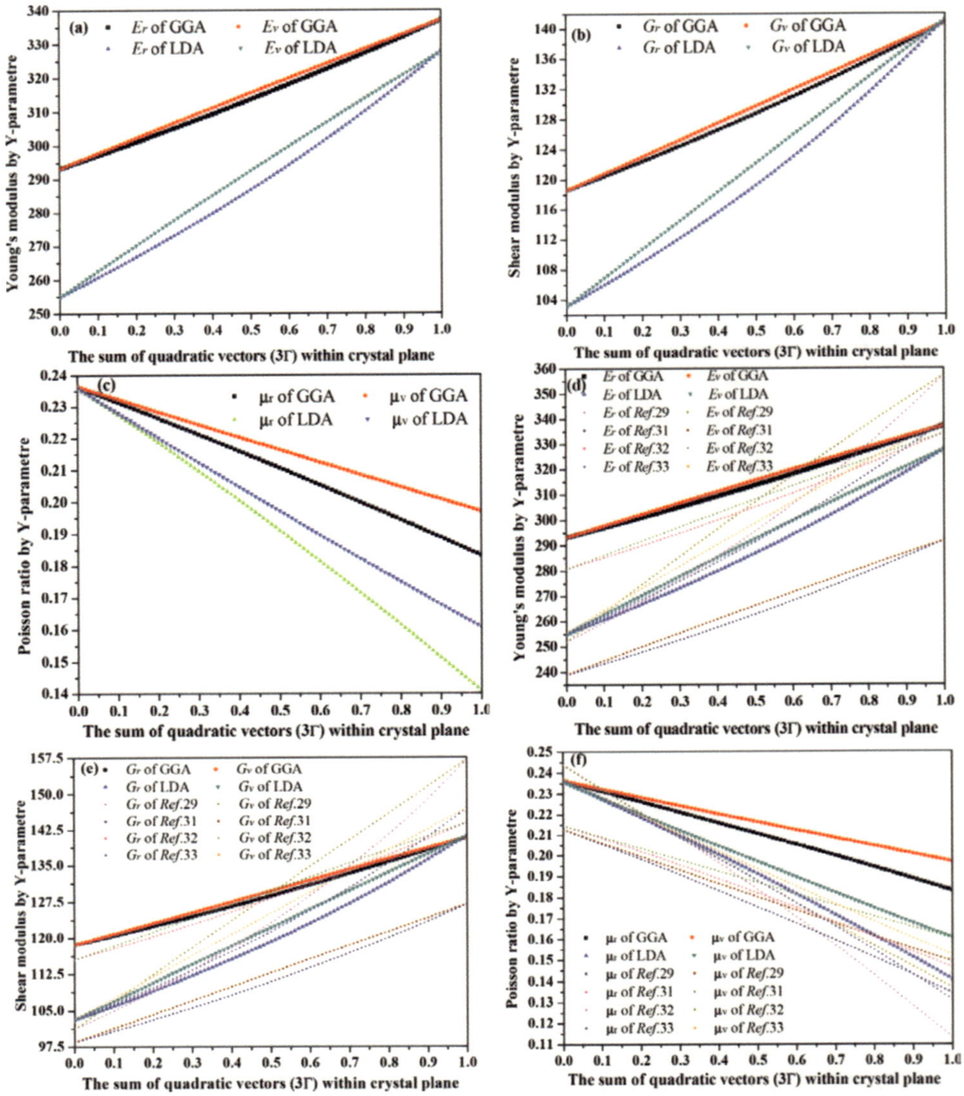

Figure 4.
Poisson ratio, Young's, and shear moduli of MgO by Y parameter.

Mechanical moduli of MgO polycrystalline by Reuss-Voigt-Hill method are shown in **Table 2**.

From **Table 2**, the MgO elastic moduli of B and G on the dependences of the elastic constants c_{ij}, the elastic anisotropic parameter A on pressure is discussed. Mechanical moduli of MgO polycrystalline by different methods are shown in **Table 2**. Elastic moduli of MgO at 0 GPa are verified and averaged as: B = 160.988 GPa, G = 124.506 GPa, and μ = 0.2360 in **Table 2**. Young's modulus of MgO is about 254.969 GPa by Reuss-Voigt-Hill estimation.

Elastic constants of MgO single crystal by GGA are as: c_{11} = 343.468 GPa, c_{12} = 106.358 GPa, and c_{44} = 140.823 GPa. By LDA, the C_{ij} are as: c_{11} = 298.508 GPa, c_{12} = 92.228 GPa, and c_{44} = 141.197 GPa. By Eqs. (26)–(28), Young's moduli of MgO at 0 GPa by GGA method in directions of *[100]*, *[110]*, and *[111]* are determined as: $E_{[100]}$ = 293.173 GPa, $E_{[110]}$ = 324.938 GPa, and $E_{[111]}$ = 337.114 GPa.

P	B_v	B_r	G_v	G_r	B	G	E	μ
Ref. [28]	138.6667	138.6667	115.6000	113.8262	138.6667	114.7131	269.7539	0.1758
Ref. [30]	164.0000	164.0000	132.6000	131.0638	164.000	131.8319	311.9172	0.1830
Ref. [31]	163.9333	163.9333	134.7860	128.7278	163.9333	131.7569	311.7504	0.1831
Ref. [32]	162.4000	162.4000	129.4800	125.6660	162.4000	127.5730	303.3001	0.1887
0 (GGA)	185.395	185.395	131.9160	130.9822	185.395	131.449	293.172	0.2364
0 (LDA)	160.988	160.988	125.974	123.037	160.988	124.506	254.969	0.2360
0.1 (GGA)	167.217	167.217	128.111	125.659	167.217	126.885	264.582	0.2363
0.2 (GGA)	165.796	165.796	128.265	125.871	165.796	127.068	264.972	0.2336
0.3 (GGA)	167.421	167.421	128.452	126.085	167.421	127.268	266.009	0.2352
0.4 (GGA)	168.718	168.718	128.673	126.362	168.718	127.517	267.279	0.2360
0.5 (GGA)	168.474	168.474	128.863	126.583	168.474	127.723	267.835	0.2350
0.6 (GGA)	168.583	168.583	129.006	126.766	168.583	127.886	268.484	0.2346
0.7 (GGA)	169.250	169.250	129.213	127.020	169.250	128.116	269.486	0.2346
0.8 (GGA)	168.603	168.603	129.409	127.258	168.603	128.334	270.054	0.2330
0.9 (GGA)	169.886	169.886	129.654	127.530	169.886	128.592	271.116	0.2340
1.0 (GGA)	170.181	170.181	129.830	127.738	170.181	128.784	271.836	0.2338

Table 2.
Mechanical moduli of MgO polycrystalline by different methods.

5. Elastic moduli of typical hexagonal polycrystals at nanoscale

The homogenized moduli of typical cubic polycrystals can be calculated by Reuss-Voigt-Hill estimation [13].

5.1 Elastic moduli of hexagonal CH structures

Elastic constants of CH single crystal at 0 GPa have been calculated by LDA calculation method and are given in Chapter 3. **Figure 5** also shows the anisotropy of the CH Young's modulus.

From **Figure 5(a)–(c)**, Y parameters of Reuss and Voigt models are very close to each other based on the elastic constants by LDA method. Besides, the Poisson ratio, Young's modulus, and shear modulus of CH by LDA method have a very good agreement with other Refs. [33–36]. Moreover, shear modulus of CH by Reuss and Voigt bounds of Y parameter is between 6.35 and 40.53 GPa under overall crystal plane orientation. Moreover, it has been found that the Young's modulus of CH is between 21.33 and 92.23 GPa in function of the crystal plane orientation. The results of Voigt and Reuss approximations show upper and lower limits in **Figure 5**; the difference arises due to different elastic constants of single CH crystals.

Mechanical moduli of CH polycrystalline by Reuss-Voigt-Hill method are shown in **Table 3**.

From **Table 3**, based on elastic constants by GGA method, elastic moduli of CH by Jia [13] are as: B = 32.41 GPa and G = 24.05 GPa. Thus, Young's modulus is averaged to 58.08 GPa by RVH method. We can see that elastic moduli by LDA method are close to other Refs. [33–36]. However, there is significant deviation with the results obtained by GGA method. Moreover, Y parameter is consistent with the

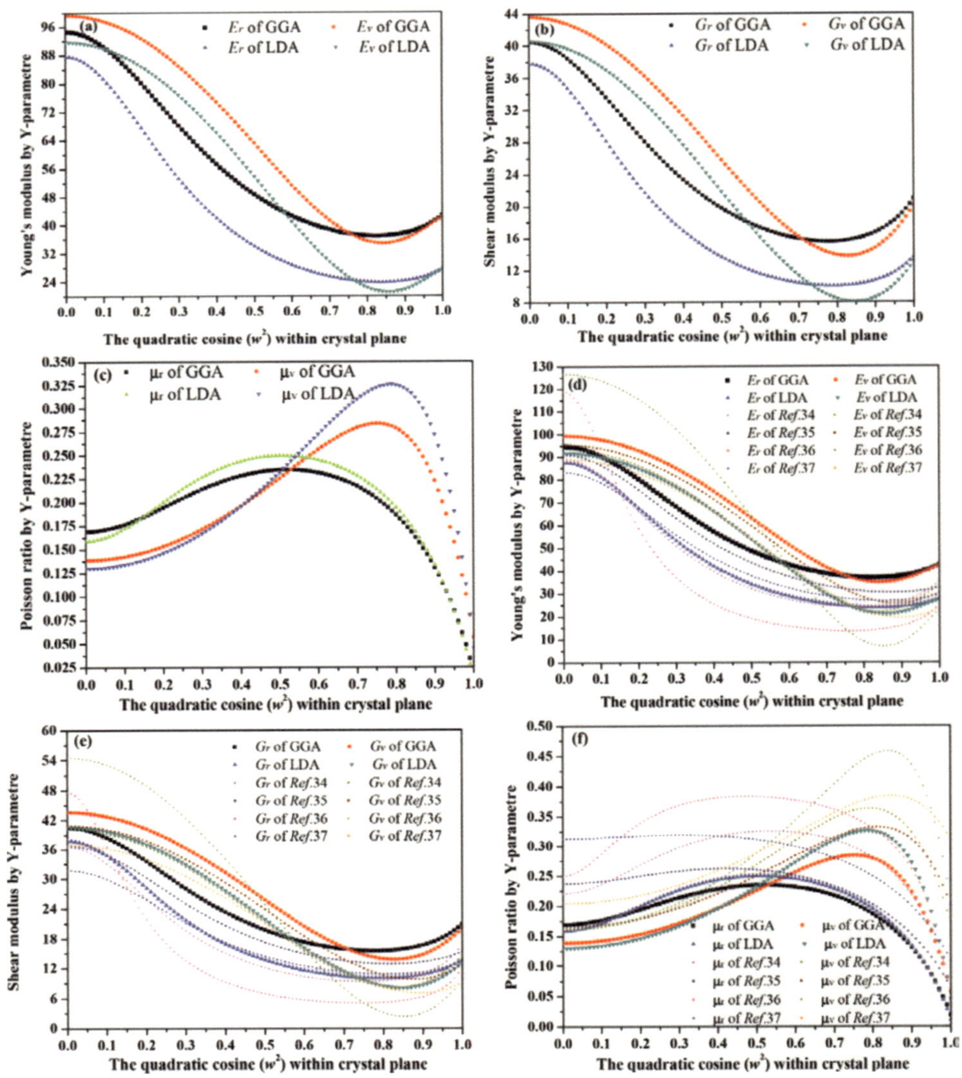

Figure 5.
Poisson ratio, Young's, and shear moduli of CH by Y parameter.

results of Voigt and Reuss models. Based on elastic constants using LDA method, by Ref. [13], elastic moduli at 0 GPa are verified and averaged as: G_v = 30.889 GPa, B_v = 22.999 GPa, G_r = 19.583 GPa, B_r = 14.890 GPa, B = 25.236 GPa, G = 18.945 GPa, and μ = 0.200 in **Table 3**. Young's modulus of CH is about 45.459 GPa by Reuss-Voigt-Hill estimation.

5.2 Elastic modulus of hexagonal calcite structure

For calcite single crystal, elastic constants have been calculated by GGA method as follows: c_{11} = 143.938 GPa, c_{12} = 55.579 GPa, c_{13} = 50.970 Pa, c_{33} = 81.175 GPa, and c_{44} = 32.658 GPa. The results of Voigt and Reuss approximations showing upper and lower limits are given in **Table 4**; the difference arises due to the use of different elastic constants of single crystals.

From **Table 4**, the results with Y parameters are consistent with the results of Voigt and Reuss models. It can also be found that E and G are satisfied with the relation $G = E/[2(1 + \mu)]$.

Pressure (GPa)	B_v (GPa)	B_r (GPa)	G_v (GPa)	G_r (GPa)	B (GPa)	G (GPa)	E (GPa)	μ
Ref. [33]	36.2300	26.6316	22.6510	13.9196	31.4308	18.2853	45.9460	0.2564
Ref. [34]	37.3 (±0.4)	26.0 (±0.3)	24.4 (±0.3)	17.5 (±0.4)	31.6 (±0.3)	20.9 (±0.3)	51.4 (±1.0)	0.23 (±0.02)
Ref. [35]	45.9778	22.3513	26.4167	8.4413	34.1646	17.4290	44.6878	0.2820
Ref. [36]	40.2422	28.0648	21.2959	15.3032	34.1535	18.2996	46.5796	0.2727
0 (GGA)	36.6395	28.1827	27.0233	21.0793	32.4111	24.0513	58.0794	0.2074
0 (LDA)	30.8885	19.5826	22.9999	14.8900	25.2355	18.9450	45.45917	0.199767
0.5 (LDA)	35.8308	27.1002	24.0796	18.2410	31.4655	21.1603	51.85653	0.225326
1.0 (LDA)	40.5673	35.0865	24.6871	17.7577	37.8269	21.2224	53.63647	0.263676
1.5 (LDA)	42.7098	35.6819	26.0083	18.9866	39.1959	22.4975	56.65329	0.259102
2.0 (LDA)	44.8341	37.4192	28.4868	25.8209	41.1267	27.1539	66.76733	0.229424
2.5 (LDA)	47.3443	41.2657	29.6026	26.3588	44.3050	27.9807	69.34408	0.239141
3.0 (LDA)	51.2366	46.0601	27.9731	23.6034	48.6484	25.7882	65.74721	0.274754
3.5 (LDA)	54.4701	49.1269	27.9306	23.5463	51.7985	25.7384	66.24323	0.286856
4.0 (LDA)	56.4897	51.9437	28.8208	25.0783	54.2167	26.9495	69.35677	0.286791
4.5 (LDA)	55.9515	51.4301	31.3143	28.8379	53.6908	30.0761	76.03140	0.263984
5.0 (LDA)	65.1811	61.3509	30.6193	26.9974	63.2660	28.8084	75.03591	0.302327

Table 3.
Mechanical moduli of CH polycrystalline by different methods.

Pressure (GPa)	B_v (GPa)	B_r (GPa)	G_v (GPa)	G_r (GPa)	B (GPa)	G (GPa)	G/B	E (GPa)	μ
Ref. [37]	80.6433	74.6632	38.9690	36.5788	77.6533	37.7739	0.486443	97.5106	0.2907
Ref. [38]	77.8078	71.5529	36.7167	35.0768	74.6803	35.8967	0.480672	92.8185	0.2929
Ref. [39]	89.7778	83.8382	39.1667	37.4911	86.8080	38.3289	0.441536	100.2343	0.3076
Ref. [40]	79.3111	72.9228	37.3967	35.1543	76.1169	36.2755	0.476576	93.9083	0.2944
Ref. [41]	78.0556	72.5157	36.5933	34.4239	75.2856	35.5086	0.471652	92.0535	0.2962
0.0 (LDA)	78.4495	70.9425	38.9105	36.1199	74.6960	37.5152	0.5022	96.4061	0.285
0.0 (GGA)	76.0099	69.6255	36.0014	28.7137	72.8177	32.3576	0.4444	84.5492	0.306
0.5 (GGA)	78.8573	72.0996	36.9119	29.1686	75.4785	33.0402	0.4377	86.4992	0.309
1.0 (GGA)	80.8540	73.9455	37.0892	29.6910	77.3998	33.3901	0.4314	87.5769	0.311
1.5 (GGA)	82.7850	76.1680	37.4184	29.7426	79.4765	33.5805	0.4225	88.3047	0.315
2.0 (GGA)	86.2955	78.5694	38.3801	30.3575	82.4325	34.3688	0.4169	90.5253	0.317
2.5 (GGA)	89.5705	80.5595	36.2223	27.4455	85.065	31.8339	0.3740	84.9097	0.334
3.0 (GGA)	90.8761	82.5141	38.9433	28.9708	86.6951	33.9571	0.3917	90.1068	0.327
3.5 (GGA)	93.6468	84.6268	39.4405	29.4758	89.1368	34.4582	0.3866	91.5743	0.329
4.0 (GGA)	95.8936	86.6201	39.1752	29.1386	91.2568	34.1569	0.3743	91.1041	0.334
4.5 (GGA)	98.4435	88.9200	39.3356	28.2723	93.6818	33.8139	0.3609	90.5476	0.306
5.0 (GGA)	100.9450	90.7940	39.7741	27.9952	95.8695	33.8846	0.3534	90.9398	0.309

Table 4.
Mechanical moduli of calcite polycrystalline by different methods.

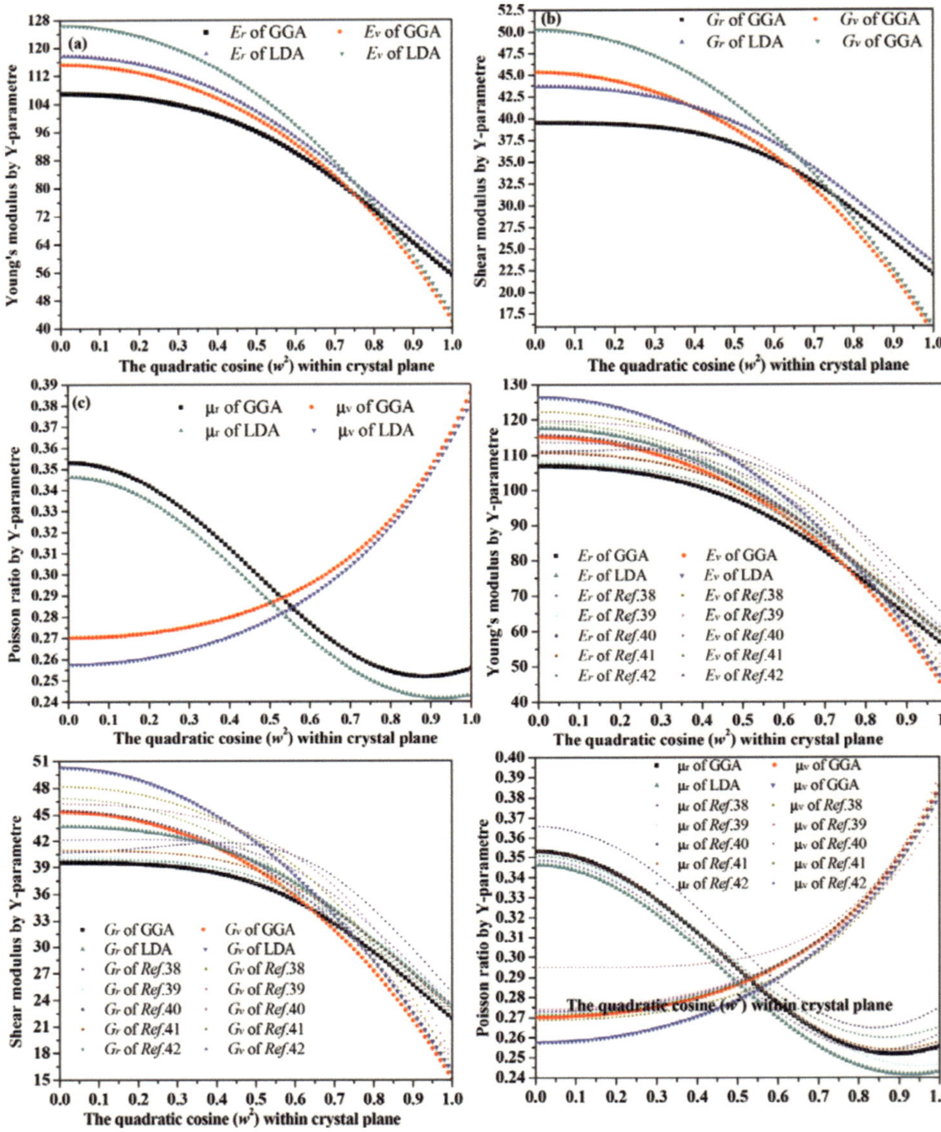

Figure 6.
Poisson ratio, Young's, and shear moduli of calcite by Y parameter.

Poisson ratio, Young's, and shear moduli of calcite structure by Y parameter are shown in **Figure 6**.

From **Figure 6(a)–(c)** of calcite polycrystals, Y parameters of Reuss and Voigt models are very close to each other based on the elastic constants by GGA method. Besides, the Poisson ratio, Young's, and shear moduli by GGA method have a very good agreement with other Refs. [37–41]. Moreover, by Reuss and Voigt bounds of Y parameter, shear modulus is between 15.08 and 46.13 GPa and Young's modulus is between 42.28 and 116.65 GPa.

Based on elastic constants by Ref. [13], elastic moduli of calcite at 0 GPa are averaged as: G_v = 36.001 GPa, B_v = 76.010 GPa, G_r = 28.714 GPa, B_r = 69.626 GPa, B = 72.818 GPa, G = 32.358 GPa, and μ = 0.306 in **Figure 6**. Young's modulus is about 84.549 GPa by RVH estimation.

Moreover, for the homogenized moduli of monoclinic polycrystals (gypsum and 11 Å tobermorite) at larger scale, the content is detailed in our previous chapter work [42]. Based on the elastic constants of 11 Å tobermorite (Ca/Si = 0.67) calculated by DFT, Young's modulus is homogenized about 79.512 GPa, which is close to the simulation result of Pellenq et al. [43] (89 GPa) and Shahsavari et al. [44] (78.939 GPa). However, Young's moduli considering the ordered Si-chain at long-range are far away from the nanoindentation test by Ulm [45, 46]. That confirms another time that the absence of order at long range in this phase and that the upscaling to polycrystals cannot be done with the tobermorite model. This finding is in agreement with the results of Manzano et al. [47].

6. Conclusions

The elastic properties of typical crystals are investigated and C_{ij} determination is given by DFT method. Y parameters have then been determined for various crystal structures and can be seen as an intermediate step in the homogenization. By means of the Y parameter, we can obtain the Young's modulus in function of the orientation of the crystal plane. Contrary to Hill approach to obtain the isotropic elastic properties of polycrystals, the Y-parameter method enables to study the anisotropic behavior of a single crystal. By means of the Y parameter, which is a function of the stiffness coefficient C_{ij} (or the compliance coefficient S_{ij}) and the crystal plane orientation, the Young's modulus in function of the orientation of the crystal plane may be obtained. Contrary to Hill approach which is used to obtain the isotropic elastic properties of a polycrystals, the Y parameter method enables to study the anisotropic behavior of a single unit cell. When $\overline{\Gamma} = 1/5$ or $w^2 = 1/3$, the result is equal to that of Hill model. Results are as follows:

1. For cubic CaO structure, Young's modulus at 0 GPa is about 175.758 GPa by elastic constants and RVH estimation. Elastic moduli are as: B = 104.894 GPa, G = 74.134 GPa, and μ = 0.2209. Young's moduli in directions of *[100]*, *[110]*, and *[111]* are separately determined by elastic constants as: $E_{[100]}$ = 175.758 GPa, $E_{[110]}$ = 181.037 GPa, and $E_{[111]}$ = 182.868 GPa.

2. For cubic MgO structure, Young's modulus of at 0 GPa is about 293.172 GPa by elastic constants and RVH estimation. Elastic moduli are homogenized as: B = 185.395 GPa, G = 131.449 GPa, and μ = 0.2364. Young's moduli at 0 GPa in directions of *[100]*, *[110]*, and *[111]* are separately determined as: $E_{[100]}$ = 293.173 GPa, $E_{[110]}$ = 324.938 GPa, and $E_{[111]}$ = 337.114 GPa.

3. For hexagonal CH structure, Young's modulus of CH at 0 GPa is 45.459 GPa by LDA and RVH methods. Other elastic moduli are as: G_v = 30.889 GPa, B_v = 23.000 GPa, G_r = 19.583 GPa, B_r = 14.890 GPa, B = 25.236 GPa, G = 18.945 GPa, and μ = 0.200. On the other hand, by GGA and RVH methods, Young's modulus is 58.07 GPa. Other elastic moduli are as: G_v = 27.023 GPa, B_v = 36.640 GPa, G_r = 21.079 GPa, B_r = 28.183 GPa, B = 32.41 GPa, G = 24.05 GPa, and μ = 0.207.

4. For hexagonal calcite structure, Young's modulus at 0 GPa is about 84.542 GPa. Elastic moduli are homogenized as: G_v = 36.001 GPa, B_v = 76.010 GPa, G_r = 28.714 GPa, B_r = 69.626 GPa, B = 72.818 GPa, G = 32.358 GPa, and μ = 0.306. Young's modulus is 84.549 GPa by RVH method.

It is worth to mention that the crystalline tobermorite structure cannot accurately represent the layered C—S—H structure at upper scale. So, modeling of a represented amorphous C—S—H structure-considered disordered Si chain, instead of the crystalline one, will be realized and further discussed based on MD method in the following Chapter 5.

Acknowledgements

The author acknowledges the financial support provided by the China Scholarship Council and the startup foundation of Xi'an Shiyou University. Thanks Qiufeng Wang for her proofreading, and very sincere thankness to prof. Fabrice Bernard and prof. Siham Kamali-Bernard for the their enthusiastic and responsible guideness.

Author details

Jia Fu[1,2]

1 Xi'an Shiyou University, Xi'an, China

2 INSA de Rennes, Rennes, France

*Address all correspondence to: fujia@xsyu.edu.cn

References

[1] Kamali-Bernard S, Bernard F. Effect of tensile cracking on diffusivity of mortar: 3D numerical modelling. Computational Materials Science. 2009; **47**(1):178-185

[2] Bernard F, Kamali-Bernard S, Prince W. 3D multi-scale modelling of mechanical behavior of sound and leached mortar. Cement and Concrete Research. 2008;**38**(4):449-458

[3] Zhu QZ, Kondo D, Shao JF. Micromechanical analysis of coupling between anisotropic damage and friction in quasi brittle materials: Role of the homogenization scheme. International Journal of Solids and Structures. 2008;**45**(5):1385-1405

[4] Luther DIT. Homogenization of damaged concrete meso-structures using representative volume elements—Implementation and application to slang [doctoral dissertation]. Weimar Germany: Bauhaus-University; 2005

[5] Zheng L, Min L. The elastic constants of polycrystalline materials with cubic system structural single crystals. Acta Physica Sinica. 2009;**58**(6):4096-4102

[6] Zheng L, Min L. Y elastic constants of polycrystalline materials with hexagonal system structure. Acta Physica Sinica. 2009;**58**(12):8511-8521

[7] Behnken H, Hauk V. Berechnung der röntgenographischen Elastizitäts-konstanten (REK) des Vielkristalls aus Einkristalldaten für beliebige Kristallsymmetrie. Zeitschrift für Metallkunde. 1986;**77**:620-626

[8] Kneer G. Die elastischen Konstanten quasiisotroper Vielkristallaggregate. Physica Status Solidi B. 1963;**3**(9): K331-K335

[9] Dixit A, Alouani M. Ab initio calculations of X-ray magnetic circular dichroism spectra within the projector augmented wave method: An implementation into the VASP code. Computer Physics Communications. 2016;**207**:136-144

[10] Gnäupel-Herold T. A software for diffraction stress factor calculations for textured materials. Powder Diffraction. 2012;**27**(02):114-116

[11] Reuss A. Berechnung der fließgrenze von mischkristallen auf grund der plastizitätsbedingung für einkristalle. ZAMM—Journal of Applied Mathematics and Mechanics/Zeitschrift für Angewandte Mathematik und Mechanik. 1929;**9**(1):49-58

[12] Voigt W. Lehrbuch der Kristallphysik Teubner; reprinted (1928) with an additional appendix. Leipzig, Teubner, New York: Johnson Reprint; 1910

[13] Fu J, Bernard F, Kamali-Bernard S. First-principles calculations of typical anisotropic cubic and hexagonal structures and homogenized moduli estimation based on the Y parameter. Application to CaO, MgO, CH and calcite $CaCO_3$. Journal of Physics and Chemistry of Solids. 2017;**101**:74-89

[14] Wang J, Li J, Yip S, Phillpot S, Wolf D. Mechanical instabilities of homogeneous crystals. Physical Review B. 1995;**52**(17):12627

[15] Karki BB, Ackland GJ, Crain J. Elastic instabilities in crystals from ab initio stress-strain relations. Journal of Physics: Condensed Matter. 1997;**9**(41):8579

[16] Morris PR. Averaging fourth-rank tensors with weight functions. Journal of Applied Physics. 1969;**40**(2):447-448

[17] Sayers CM. Ultrasonic velocities in anisotropic polycrystalline aggregates.

Journal of Physics D: Applied Physics. 1982;**15**(11):2157-2167

[18] Li Y, Thompson RB. Relations between elastic constants C_{ij} and texture parameters for hexagonal materials. Journal of Applied Physics. 1990;**67**(5): 2663-2665

[19] Anderson AJ, Thompson RB, Cook CS. Ultrasonic measurement of the Keams texture factors inzirealoy, zirconium, and titanium. Metallurgical and Materials Transactions. 1999;**30A**:1981-1988

[20] Anderson OL. A simplified method for calculating the Debye temperature from elastic constants. Journal of Physics and Chemistry of Solids. 1963; **24**(7):909-917

[21] Hill R. The elastic behaviour of a crystalline aggregate. Proceedings of the Physical Society. Section A. 1952; **65**(5):349

[22] Sokolnikoff IS, Specht RD. Mathematical Theory of Elasticity. Vol. 83. New York: McGraw-Hill; 1956

[23] The Society of Materials Science, Japan. X-Ray Stres Measurement. Tokyo: Yokendo; 1990. pp. 4-20 (in Japanese)

[24] Bergersen B. Physics of Materials Material. University of British Columbia. pp. 47-49. http://www.phas.ubc.ca/~birger/315out.htm

[25] Onami M, Ivasimidzu S, Genka K, Siodzava K, Tanaka K. Introduction to Micromechanics. Moscow: Metallurgiya; 1987

[26] Tanaka K, Matsui E, Kurimura T, Akiniwa Y. Elastic constants of sintered alumina for X-ray stress measurements. Journal of the Society of Materials Science, Japan. 1987;**36**(407):792-798

[27] Mehl MJ, Hemley RJ, Boyer LL. Potential-induced breathing model for the elastic moduli and high-pressure behavior of the cubic alkaline-earth oxides. Physical Review B. 1986;**33**(12):8685

[28] Karki BB, Crain J. Structure and elasticity of CaO at high pressure. Journal of Geophysical Research: Solid Earth (1978–2012). 1998;**103**(B6): 12405-12411

[29] Tsuchiya T, Kawamura K. Systematics of elasticity: Ab initio study in B1-type alkaline earth oxides. Journal of Chemical Physics. 2001;**114**(22): 10086-10093

[30] Fiquet G, Richet P, Montagnac G. High-temperature thermal expansion of lime, periclase, corundum and spinel. Physics and Chemistry of Minerals. 1999;**27**(2):103-111

[31] Sinogeikin SV, Bass JD. Single-crystal elasticity of pyrope and MgO to 20 GPa by Brillouin scattering in the diamond cell. Physics of the Earth and Planetary Interiors. 2000;**120**(1):43-62

[32] Matsui M, Parker SC, Leslie M. The MD simulation of the equation of state of MgO: Application as a pressure calibration standard at high temperature and high pressure. American Mineralogist. 2000;**85**(2):312-316

[33] Laugesen JL. Density functional calculations of elastic properties of portlandite, $Ca(OH)_2$. Cement and Concrete Research. 2005;**35**(2):199-202

[34] Speziale S, Reichmann HJ, Schilling FR, Wenk HR, et al. Determination of the elastic constants of portlandite by Brillouin spectroscopy. Cement and Concrete Research. 2008;**38**(10): 1148-1153

[35] Kerisit S, Parker SC, Harding JH. Atomistic simulation of the dissociative adsorption of water on calcite surfaces. The Journal of Physical Chemistry B. 2003;**107**(31):7676-7682

[36] Holuj F, Drozdowski M, Czajkowski M. Brillouin spectrum of $Ca(OH)_2$. Solid State Communications. 1985;**56**(12): 1019-1021

[37] Aaeid A. Études ab initio et dynamique moléculaire des propriétés structurales et thermodynamiques de la calcite et la witherite sous hautes pressions [PhD Thesis]. France: l'Université des Sciences et Technologies de Lille (In French); 2010

[38] Dandekar DP. Pressure dependence of the elastic constants of calcite. Physical Review. 1968;**172**(3):873-877

[39] Le Page Y, Saxe P, Rodgers JR. Ab initio stiffness for low quartz and calcite. Physica Status Solidi B. 2002; **229**(3):1155-1161

[40] Chen CC, Lin CC, Liu LG, et al. Letters. Elasticity of single-crystal calcite and rhodochrosite by Brillouin spectroscopy. American Mineralogist. 2001;**86**(11–12):1525-1529

[41] Dandekar DP, Ruoff AL. Temperature dependence of the elastic constants of calcite between 160 and 300 K. Journal of Applied Physics. 1968; **39**(13):6004-6009

[42] Jia F. Elastic constants and homogenized moduli of monoclinic structures based on density functional theory. In: Yang G, editor. Density Functional Calculations: Recent Progresses of Theory and Application. Stafa-Zurich: Trans Tech Publications; 2018

[43] Pellenq RJ-M, Nicolas Lequeux, Henri Van Damme, Engineering the bonding scheme in C-S-H: the iono-covalent framework. Cement and Concrete Research. 2008;**38**(2):159-174

[44] Shahsavari R, Buehler MJ, Pellenq RJM, et al. First-principles study of elastic constants and interlayer interactions of complex hydrated oxides: Case study of tobermorite and jennite. Journal of the American Ceramic Society. 2009;**92**(10):2323-2330

[45] Vandamme M, Ulm FJ. Nanoindentation investigation of creep properties of calcium silicate hydrates. Cement and Concrete Research. 2013; **52**:38-52

[46] Miller M, Bobko C, Vandamme M, Ulm F-J. Surface roughness criteria for cement paste nanoindentation. Cement and Concrete Research. 2008;**38**: 467-476

[47] Manzano H, Dolado JS, Guerrero A, et al. Mechanical properties of crystalline calcium-silicate-hydrates: Comparison with cementitious C—S—H gels. Physica Status Solidi. 2007;**204**(6):1775-1780

Chapter 5

Classical Molecular Dynamics (MD): Atomistic Simulation of Typical C—S—H Structures

Jia Fu

Abstract

C—S—H has a great influence on the mechanical properties of cement and it is the most important binding phase of cement paste. In the light of recent computational material technology, a promising way to establish a larger microscale structure to obtain the desired C—S—H phase structure with the consideration of porosity is proposed, using a mechanical property to verify the rationality of the designed structure. Here, we discuss the basic atomic unit at nanoscale and the inverse approach, which propose strategies for the design of a possible C—S—H structure over nano- to macroscales of an achievement of nearly experimental phase. At the nanoscale, a 11 Å tobermorite monoclinic crystal with the ordered Si-chain is used to enlarge the scale to the monolithic "globule" C—S—H structure about 5.5 nm^3 using molecular dynamics simulation. However, the inverse approach with full structural C—S—H information (atoms and their positions) from the macroscale by the neutral scattering tests to nanoscale by the reorganized amorphous C—S—H cell is unlinked nowadays. Our contribution is to find the interlinkages between the "globule" at nanoscale with LD/HD C—S—H at microscale, thus to seek experimental validation of the C—S—H hydrated phases.

Keywords: nanoscale, tobermorite, molecular dynamics, DFT, anisotropic elasticity, Young's moduli

1. Introduction

Concrete is the most used building material in the world, because of its appreciable advantages such as load-bearing capacity, durability, maneuverability, flowability. Recently, a platform called multiscale modeling of computational concrete (MuMoCC) using various modeling methods has been developed. In these previous works, cementitious materials have been studied at four different scales:

i. The nanoscale that considers the element cells of the various components of the hardened cement paste [1–3]. The anisotropic elastic properties of different hydrated phases are determined thanks to various modeling methods [4]: density functional theory, molecular dynamics, atomic finite element method.

ii. The microscale that considers the hardened cement paste as a heterogeneous and complex porous material in which the main solid phases are calcium silicate hydrates (C—S—H), portlandite (calcium hydroxides or CH), and hydrated aluminates or sulfoaluminates (ettringite or Aft, monosulfoaluminate or AFm) phases. The solid phase is itself in equilibrium with the internal solution, which fills the porosity [5].

iii. The mesoscale that is in fact divided into two scales: the submesoscale, firstly, is the scale of mortar, that is, the smallest fraction of the particle size distribution bonded by homogeneous cement paste with porosity and air voids; then the mesoscale itself is the scale of concrete: the largest aggregates embedded in a matrix of mortar. These two scales introduce also the presence of the interfacial transition zone between particles and matrix, which modifies crack patterns and diffusion paths but not necessarily homogeneous effective physical and mechanical properties at the various scales [6–8].

In such a hierarchical approach, the adequate outputs at a given scale are used as input data at the next higher level.

C—S—H as the main and typical constituents of hydrated cementitious systems separately represents approximately 70% of the volume fraction of portland cement paste [9], which has a great influence on the mechanical properties of the cement paste, and it is the most important binding phase of cement paste. Its Young's moduli are needed in the modeling of cement at the macro- and microscales [10]. The traditional continuum models [11] and nonlocal continuum theory [12] are not adequate in modeling [13] of these materials [14]. As C—S—H theoretical mechanism is very complex [15] and size effect [16, 17] are not commonly considered, the simulation is used for comparison of mechanical properties of C—S—H according to loading/unloading curves between load and displacement in nanoindentation technique. Although there are some relative reports on experimental analysis and the numerical calculation of C—S—H structure, its nanostructure has not yet been revealed.

The crystal structure of C—S—H structure is basically known and commonly modeled as tobermorite-like (i.e., tobermorite 9, 11, 14 Å) and jennite-like systems [18] and/or with distorted semicrystalline variations of them [19]. Tobermorite as one of the earliest models proposed by Taylor and Howison [20] is thought as its hydration degree, which may describe the relative C—S—H nanostructure. From the view of Ca/Si ratio, there are two types of C—S—H structures: the C—S—H(I) (with Ca/Si of 0.6–1.5) and C—S—H(II) structures [21] (with Ca/Si of 1.5–2.0), of which the later is close to the experimentally confirmed structure $C_{1.7}$—S—$H_{1.8}$ with Ca/Si of 1.7 in the C—S—H gel [22]. From the view of several nanometers length scale, C—S—H gel has a nanogranular aspect composed of monolithic C—S—H (full-dense C—S—H) and porosity. At this scale, C—S—H gel "globule" model exists in two forms: low-density C—S—H (LD C—S—H) and high-density C—S—H (HD C—S—H). Recently, Fabrice et al. have investigated the specific heat of cementitious materials using multiscale modeling approach [23]. Hou et al. [24] have investigated an amorphous C—S—H structure on the elastic properties of the layered C—S—H based on the CSH-FF field.

In recent years, some scholars have successfully simulated C—S—H structure by MD simulations. Pellenq et al. [25] have simulated several tobermorite structures by using the General Utility Lattice Program (GULP), where GULP uses the core-shell potential. Kalinichev et al. [26] have simulated the tobermorite structure using ClayFF force field [27], where water molecules are modeled using simple point charge (SPC) model, and it is indicated that the water shows very strong binding force in the solid surface. Pellenq et al. [28] have developed a new modeling method to simulate C—S—H gel structure, of which it has a shorter $[SiO_4]^{4-}$ chain under the system equilibrium structure $C_{1.65}$—S—$H_{1.75}$, comparing with experimental values of $C_{1.7}$—S—$H_{1.8}$ [22, 29]. Shahsavari et al. [30] proposed a new CSH-FF field with a higher accuracy and more efficient than the traditional core-shell potential. Ji et al. [31] have simulated the influencing role of a variety of water molecules (usually optional SPC model) on C—S—H by using GROMACS [32] program and mainly used a simple harmonic force potential description of C—S—H between atoms and ions. Zaoui [33] has simulated tobermorite structure of different Ca/Si ratios under various pressures, of which Buckingham potential and core-shell potential are used. This author found that the elastic modulus of different Ca/Si ratio tobermorite structures tends to be similar with increasing pressure. Some scholars [34] develop the DL_POLY program to simulate the tobermorite by MD, finding that amorphous form of C—S—H(I) is a short-range structure and long-range disordered structure, which is more consistent with the experimental parameters. Dai et al. [35] have compared two C—S—H structures of different Ca/Si ratios and discussed MD modeling parameters, of which the COMPASS force field is used. According to the NMR measurement of C—S—H [19], mechanical properties of this amorphous structure can be simulated by MD method to study the deformation mechanism at atomic scale.

Moreover, microporomechanics technique has been used to calculate elastic properties of LD C—S—H and HD C—S—H gels [36]. For the elastic properties of LD and HD C—S—H, Constantinides and Ulm [37] have obtained elastic moduli by nanoindentation experiment. The porosity is also a factor of influencing elastic modulus of cement paste, and it has been investigated based on the backscattered electron image analysis and the HYMOSTRUC model [38]. The impact of the porosity on the matrix Young's modulus is that the Young's modulus decreases highly with the increasing of the porosity [39]. According to LD and HD C—S—H models described by Jennings [22], the gel porosity of LD C—S—H solid phase is 35–37%, while the gel porosity of HD C—S—H solid phase is 24%. However, the relationship between the C—S—H structure with the size of about 5 nm and C—S—H phases (LD and HD C—S—H) has not been revealed yet. In this chapter, the porosity and homogenization are used to explain the difference, which is meaningful for the verification of both nanoindentation simulation [40] and nanoindentation experiment [41, 42] at nanoscale in Chapter 6.

2. Molecular dynamics modeling of typical C—S—H structures by LAMMPS

Molecular dynamics is a deterministic method that offers the possibility of a microscopic description of a physical system in consideration of all the interactions involved. The main advantage of this method is that it gives the information on the evolution of the system over time and this by numerically solving Hamilton equations of motion, Lagrange or Newton.

2.1 Principle of MD method and ClayFF field used in LAMMPS

2.1.1 Introduction and principle of molecular dynamics method

The applied numerical integration for Newton's equation of motion by MD [43] is described as:

$$H = \frac{1}{2}\sum_{i=1}^{N}\frac{p_i^2}{m_i} + \sum_{i=1}^{N-1}\sum_{j=i+1}^{N} U(r_{ij}) \tag{1}$$

$$\mathbf{p}_i = m_i\frac{d\mathbf{r}_i}{dt} = m_i\mathbf{v}_i \tag{2}$$

$$\frac{d\mathbf{p}_i}{dt} = \sum_{i=1}^{N-1}\sum_{j=i+1}^{N}\mathbf{F}(r_{ij}) = -\sum_{i=1}^{N-1}\sum_{j=i+1}^{N}\frac{\partial U(r_{ij})}{\partial \mathbf{r_{ij}}} \tag{3}$$

where p is momentum that depends on the mass m and the acceleration a, F is the net force vector on the ith molecule, r is the position vector of the ith particle. The force on each molecule is obtained from the interaction potential, that is, the Lennard-Jones pair potential.

Positions and velocities of the particles are computed by integrating the equations of motion using finite difference algorithm at equal time intervals. One of the most used algorithms are the ones developed by Verlet that is the velocity Verlet leapfrog algorithm, which makes use of the half-time step velocities to calculate positions and velocities of particles [44]:

$$\mathbf{r_i}(\mathrm{t} + \Delta t) = \mathbf{r_i}(\mathrm{t}) + v_\mathrm{i}(\mathrm{t} + \Delta t/2)\,\Delta t \tag{4}$$

$$\mathbf{v_i}(\mathrm{t} + \Delta t/2) = \mathbf{v_i}(\mathrm{t}\text{-}\Delta t/2) + \frac{d^2\mathbf{r}_i(t)}{dt^2}\Delta t + \mathrm{O}(\Delta t^3) \tag{5}$$

$$\frac{d^2\mathbf{r}_i(t)}{dt^2} = \frac{\mathbf{F_i}}{m_\mathrm{i}} = -\frac{1}{m_\mathrm{i}}\sum\nabla\cdot\mathrm{U(r)} \tag{6}$$

where F is the net force vector on the ith molecule, m is the mass, and r is the position vector of the ith particle. For Verlet algorithm, particle acceleration: $\mathbf{a_i}(\mathrm{t}) = \frac{\mathbf{F_i}(\mathrm{t})}{m_\mathrm{i}}$. Initial conditions of Verlet algorithm are as: $\mathbf{r}_i|_{t=0} = \mathbf{r}_i(0)$, $\frac{d\mathbf{r}_i}{dt}|_{t=0} = \mathbf{v}_i(0)$.

There are a few methods to control temperature, the simplest method involves velocity scaling or coupling with heat bath, and other more complicated approaches are the Andersen and Nose-Hoover thermostats [45]. In this research, a special form of Nose-Hoover thermostats developed by Berendsen is used, in which the friction coefficient λ rather than its time derivative varies according to the following equation rather than its derivative [46]:

$$\lambda = \left(\sum_{i=1}^{N}\frac{p_i^2}{2m_i} - \frac{gkT}{2}\right)/Q \tag{7}$$

where $Q = gkT\tau^2$, g is the degree of freedom in the system ($g = 3N - 4$) for a thermostated monatomic system with zero momenta, τ is thermostat relaxation time typically about 0.5.

From **Figure 1**, the domain reduction only performs analysis of a representative substructure. During MD simulation, only one box is modeled explicitly. Each particle interacts with other particles in the box and with their images in nearby boxes. The mean square displacement (MSD) is a measure of the average distance a molecule travels, it represents the statistical average of the particle trajectories change over time. By solving differential motion equations, which is as [44]:

$$MSD(t) = \frac{1}{n}\sum_{i=1}^{n}\left\langle |r_n(t) - r_n(o)|^2\right\rangle \tag{8}$$

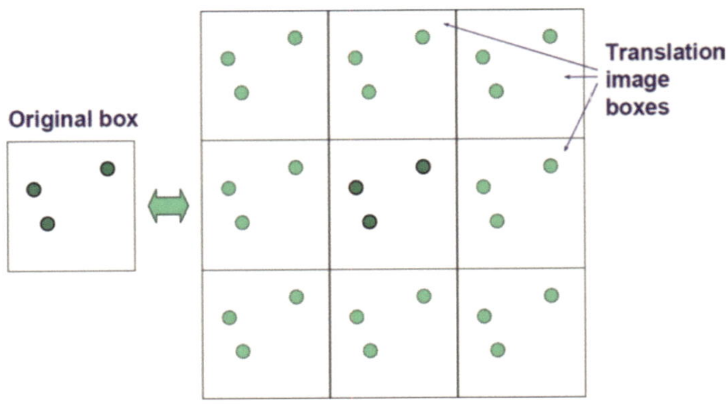

Figure 1.
Diagram from the original box to the conversion box in statistical mechanics.

where $r_n(t)$ is the atom position at time t.

The diffusion coefficient is one-sixth of the slope of the mean square displacement, in **Figure 2**.

As shown in **Figure 2**, the probability to find a pair of a distance r apart, relative to a uniform random distribution of particles at the same density, can be defined as $g(r)$. For each particle, the probability to find another particle within the shell r to $r + \Delta r$ is measured by $g(r)$. Radial distribution function (RDF) is a measure of arrangement of atoms in a structure, and the MD is a deterministic method that offers the possibility of a microscopic description of a physical system in consideration of all the interactions involved. In simulation, an input file was generated for LAMMPS code to calculate the response of C—S—H structures with a large number of particles (atoms) to uniaxial loading, which leads to strain-stress data to calculate elastic modulus.

2.1.2 Parameters of ClayFF field and programming code by LAMMPS

The total potential energy of the classical molecular force field expands to:

$$u = \sum_{bonds} u_s + \sum_{angles} u_b + \sum_{torsions} u_t + \sum_{oops} u_o + \sum_{improoers} u_i + \sum_{inv} u_{inv} + \sum_{cross} u_{cross} + \sum_{vdW} u_{vdW} + \sum_{cl} u_{cl} \tag{9}$$

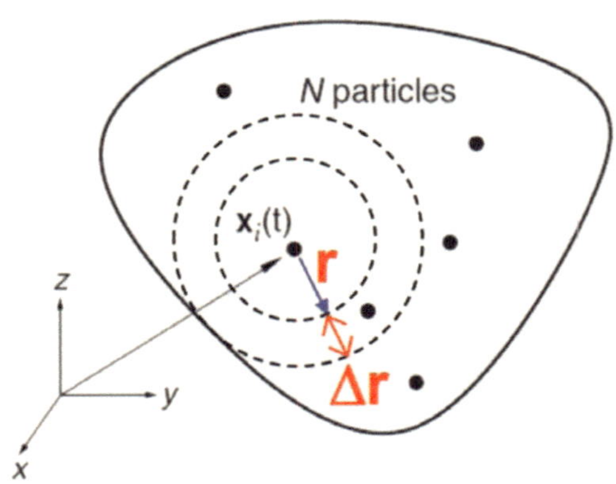

Figure 2.
Description of the probability to find a particle within the shell r to r + Δr *[44].*

The total energy is the sum of Coulombic (electrostatic) interactions, short-range interactions (van der Waals), and bonded (stretching/angular) interactions [27]:

$$U_{total} = U_{vdw} + U_{coul} + U_{bond} + U_{angle} \tag{10}$$

For bonded intramolecular interactions, the Coulombic and VDW interactions are excluded. The Coulombic energy is represented as:

$$U_{coul} = \frac{e^2}{4\pi\varepsilon_0} \sum_{i<j} \frac{q_i q_j}{r_{ij}} \tag{11}$$

where q_i and q_j are partial charges, e is the charge of the electron, while ε_0 is the dielectric permittivity of vacuum (8.85419×10^{-12} F/m). Note that the Coulombic interaction is long range and requires techniques such as the Ewald sums to be properly calculated. The VDW interactions are represented with the conventional 12-6 Lennard-Jones function that includes the short-range repulsion and the attractive dispersion energy:

$$U_{vdw} = \sum_{i<j} D_{ij} \left[\left(\frac{R_{ij}}{r_{ij}} \right)^{12} - 2 \left(\frac{R_{ij}}{r_{ij}} \right)^{6} \right] \tag{12}$$

where D and R_{ij} are empirical parameters derived from the fitting of the ClayFF model to a number of observed structural property data for oxides, hydroxides, and oxyhydroxides.

The interaction parameters between the unlike atoms are calculated by the arithmetic mean rule for the distance parameter R_{ij}, the geometric mean rule for the energy parameter D_{ij}:

$$R_{ij} = \frac{1}{2}\left(R_i + R_j\right) \text{ and } D_{ij} = \sqrt{D_i D_j} \tag{13}$$

Bond stretching energy is considered between O and H of either a hydroxyl or a water molecule and is described by a simple harmonic term as:

$$U_{bond} = k_1 \left(r_{ij} - r_0\right)^2 \tag{14}$$

where k_1 is a force constant and r_0 represents the equilibrium bond length, both values taken from the flexible version of the SPC water model. For the description of the vibrational motion of hydroxyl groups, a bending (three-body) term is introduced in form of a harmonic relationship:

$$U_{angle} = k_2 \left(\theta_{ijk} - \theta_0\right)^2 \tag{15}$$

where k_2 is a force constant of Eq. (15), θ_{ijk} is the bond angle for the oxygen-hydrogen, and θ_0 refers to the equilibrium bond angle.

2.1.3 Parameters of ClayFF field used in C—S—H structures

In this study, bonded interactions include bond stretching potential energy function and bond angle bending potential energy function, which can be described by a harmonic potential function [47]. For nonbonded parameters of

ClayFF force field, Lennard-Jones (12-6) potential is mainly considered. The empirical interatomic potentials in ClayFF field have been described above, and the corresponding parameters used for C—S—H structures are listed in **Tables 1–3**.

In fact, development of new parameters is very difficult, we mainly look for force field and select suitable parameters to be programmed in LAMMPS software, as many existed and complex force fields at present are difficult to be fully taken into account in this developing software.

2.2 Modeling of typical C—S—H structures considering the Ca/Si ratio

2.2.1 Description of C—S—H structures with various Ca/Si and Qn ratios

In neat Portland cement, only the C—S—H with the highest Ca/Si ratio (>~1.5) is observed, whereas C—S—H with the whole compositional range may exist in

(a) Bond stretching potential parameters				
Species *i*	**Species *j***		**K_1 (kcal mol^{-1} Å^{-2})**	**R_0 (Å)**
ow	hw		554.1349	1.0000
oh	ho		554.1349	1.0000
ohs	ho		554.1349	1.0000
(b) Bond angle bending potential parameters				
Species *i*	**Species *j***	**Species *k***	**K_1 (kcal mol^{-1} rad^{-2})**	**θ_0 (degree)**
hw	ow	hw	45.7696	109.4700

Table 1.
Bonded parameters of ClayFF force field [27].

Type	R_0 (Å)	D_0 (kcal/mol)	Element	Charge ($\bar{e}$)	Atom type specification
h*	4.5775	0.0000	H	0.410	Water hydrogen
ho	4.5775	0.0000	H	0.425	Hydroxyl hydrogen
o*	3.5532	0.1554	O	−0.820	Water oxygen
oh	3.5532	0.1554	O	−0.950	Hydroxyl oxygen
ob	3.5532	0.1554	O	−1.050	Bridging oxygen
St	3.7064	1.8405E−06	Si	2.100	Tetrahedral silicon
cao	6.2484	5.0298E−06	Ca	1.360	Octahedral calcium
obss	3.5532	0.1554	O	−1.300	Bridging oxygen with double substitution
obts	3.5532	0,1554	O	−1.169	Bridging oxygen with tetrahedral substitution
obos	35,532	0.1554	O	−1.181	Bridging oxygen with octahedral substitution
ohs	3.5532	0.1554	O	−1.081	Hydroxyl oxygen with substitution
cah	6.2428	5.0298E−06	Ca	1.050	Hydroxide calcium
oh-	3.5532	0.1554	O	−1.410	Oxygen in hydroxyl anion
Ca	3.2237	0.1000	Ca	2.000	Aqueous calcium ion

Table 2.
The nonbonded parameters of ClayFF force field [27].

(a) Charge and species of atoms of C—S—H (C/S = 1.67) structure used [27]			
Species	**Charge ($\bar{e}$)**	**D_0 (kcal/mol)**	**R_0 (Å)**
Water hydrogen (Hw)	0.41	—	—
Hydroxyl hydrogen (Ho)	0.42	—	—
Water oxygen (Ow)	−0.82	0.1554	3.5532
Hydroxyl oxygen (Oh)	−0.95	0.1554	3.5532
Bridging oxygen (O)	−1.05	0.1554	3.5532
Silicon (Si)	2.1	1.84E−06	3.7064
Calcium (Ca)	1.05	5.03E−06	6.2428

(b) The nonbonded Lennard-Jones (12-6) of C—S—H (C/S = 1.67) structure [27]							
Species *i*	**Species *j***	**D_{ij}**	**R_{ij}**	**Species *i***	**Species *j***	**D_{ij}**	**R_{ij}**
Ca	Ca	5.0298E−06	6.2484	Sz	Ow	0.000534802	3.6298
Ca	Cw	5.0298E−06	6.2456	Sz	Sz	1.8405E−06	3.7064
Ca	Sz	3.04259E−06	4.9774	Sz	O	0.000534802	3.6298
Ca	Oss	0.000884099	4.9008	Sz	Oss	0.000534802	3.6298
Ca	O	0.000884099	4.9008	Oss	O	0.1554	3.5532
Ca	Hw	0	3.1242	Oss	Oss	0.1554	3.5532
Ca	Ow	0.000884099	4.9008	Oss	Hw	0	1.7766
Cw	Hw	0	3.1214	Oss	Ow	0.1554	3.5532
Cw	Ow	0.000884099	4.898	O	O	0.1554	3.5532
Cw	Sz	3.04259E−06	4.9746	O	Ow	0.1554	3.5532
Cw	Cw	5.0298E−06	6.2428	O	Hw	0	1.7766
Cw	O	0.000884099	4.898	Ow	Ow	0.1554	3.5532
Cw	Oss	0.000884099	4.898	Ow	Hw	0	1.7766
Sz	Hw	0	1.8532	Hw	Hw	0	0

Table 3.
Parameters for ClayFF field of C—S—H (C/S = 1.67) structure used [27].

cement pastes containing fly ash, metakaolin, or silica fume [48]. From the view of several nanometers length scale, C—S—H gel has a nanogranular aspect composed of monolithic C—S—H (full-dense C—S—H) and porosity. At his scale, C—S—H gel exists in two forms: low-density C—S—H and high-density C—S—H. Two typical C—S—H structures of LD and HD C—S—H have been provided by Jennings [49] and Masoero et al. [50] in **Figure 3**.

From **Figure 3**, from SEM image of C—S—H, two typical structures at nanoscale are provided, of which the LD C—S—H solid phase with the diameter of C—S—H gel is less than 16.6 nm, the gel porosity is about 35–37%, while the gel porosity of HD C—S—H solid phase is 24%. The solid porosity manifests itself (about 18%) at a scale smaller than the characteristic solid dimension of 2.2 nm [49, 50], irrespective of the type of C—S—H. It should be noted that, beyond this scale of nanoporosity, there is another type of porosity, which was found to differ from two C—S—H structures. This solid phase is named “globules,” which was found to have a characteristic size of 5.6 nm.

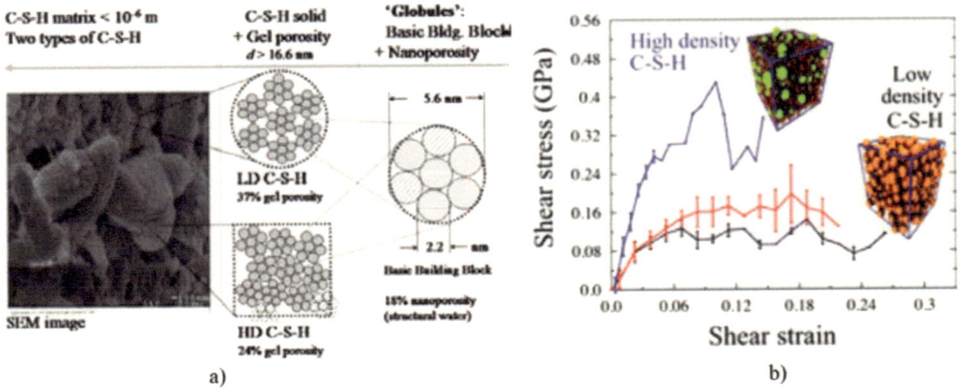

Figure 3.
Structures at nanoscale and shear moduli of two typical C—S—H [49, 50]. (a) Two typical C—S—H structures at nanoscale [49] and (b) Shear moduli of typical LD and HD C—S—H structures [50].

Relation of C—S—H phase at microscale and "globule" C—S—H at nanoscale is linked, in **Figure 4**.

From **Figure 4**, the porosity is an important factor to reveal the differences between "globule" C—S—H at nanoscale and LD/HD C—S—H phase at microscale. Then elastic modulus of the "globule" C—S—H about 5.5 nm^3 can be adopted to assess the elastic moduli of LD and HD C—S—H phases, where the values are verified by the outer and inner C—S—H distinguished in nanoindentation experiment.

It is generally considered that C—S—H solid solution structure is a kind of tobermorite structure between CH structures to form a sandwich structure [51]. Calcium-rich model and silicon-rich structural models: depending on the Ca/Si ratio, C—S—H can be divided into silicon-rich C—S—H (Ca/Si = 0.65–1.0) and calcium-rich C—S—H (Ca/Si = 1.1–1.7) [52]. Structure and defect model of C—S—H [53] and the NMR measurement of Grutzeck et al. [19, 52], $Q_1/Q_2 \sim C/(S + A)$ are as in **Figure 5**.

From **Figure 5(a)**, we can see that such nanostructures and intermediary structural models consider that C—S—H composition is uniform within the scope of nanocrystals (about 5 nm). However, in the range of short-range order of 1 nm, C—S—H composition and structure can vary, especially in the area of matrix

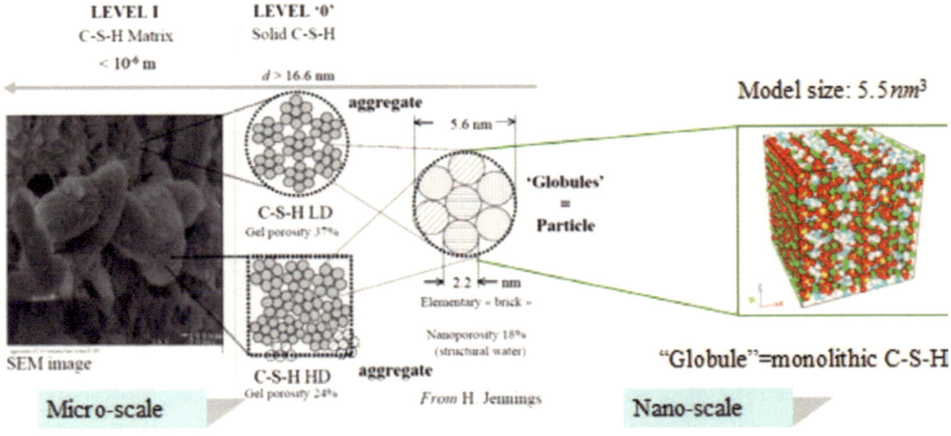

Figure 4.
Relation of C—S—H phase at microscale and "globule" C—S—H at nanoscale.

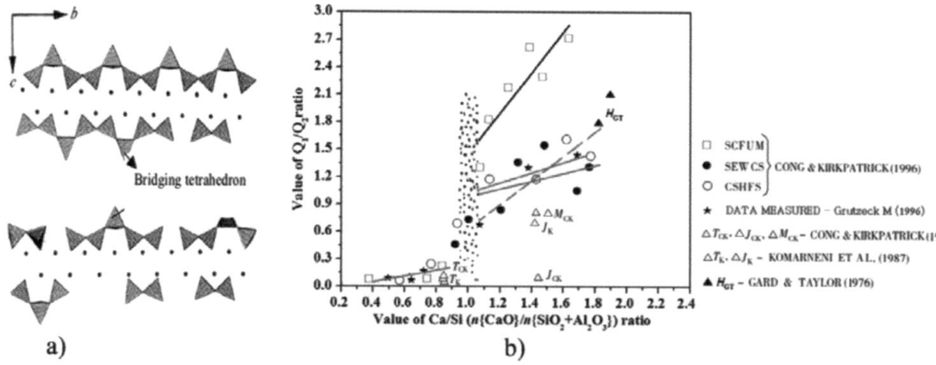

Figure 5.
Structure and defect model of C—S—H and the NMR measurement data of Q_1/Q_2 ratio. (a) Structure and defect model of C—S—H [53] and (b) Q_1/Q_2 ratio of various C—S—H [19, 52].

composition of the amorphous C—S—H. From **Figure 5(b)**, the results show that the measured Ca/Si ratio has a mutation in both sides of 1.0–1.1 region, which proved that C—S—H between 1.0 and 1.1 is vacant. However, Q_1/Q_2 in calcium-rich C—S—H is fluctuated from 1.0 to 1.5, which can be considered about 50% dimers and 50% repeated silicon-oxygen tetrahedron chain ternary.

There, a Q_n nomenclature is used in general for the peaks. Q_n is the chemical shift of a silicon atom, which is bound to n bridging oxygens. The Q_n distribution and the mean chain lengths (MCL) of C—S—H structures with various Ca/Si ratios are given by Tajuelo et al. [54], shown in **Table 4**.

From **Table 4**, when the C—S—H structure with a C/S ratio higher than 1.0, the number of Q_3 is zero, which means there is no silicate tetrahedral with three bridging oxygen atoms in its structure.

2.2.2 Structural, bonding, and interaction of C—S—H structures

For modeling of a 14Å tobermorite: (C/S = 0.83), we chose Bonaccorsi model [55] with the empirical formula of $Ca_5Si_6O_{16}(OH)_2{\cdot}7H_2O$ (Ca/Si = 0.83), and crystallographic parameters are as follows: (1) crystal system: triclinic; (2) the cell parameters: a = 6.735, b = 7.425, c = 27.987 Å, α = 90°, β = 90°, γ = 123.25°; (3) space group: *B11b*. **Figure 6** shows the structure and atomic connections of 14 Å tobermorite.

From **Figure 6**, 14 Å tobermorite structure is basically a layered structure. Besides, the central part is a Ca—O sheet (with an empirical formula: CaO_2,

Ca/Si	Q_1 (%)	Q_2 (%)	Q_3 (%)	MCL
0.75	9.5	79.8	10.7	18.8
0.83	19.1	77.1	3.8	10.1
1	35.7	62.0	2.3	5.5
1.25	74.0	26.0	0	2.7
1.33	79.9	20.1	0	2.5
1.5	87.7	12.3	0	2.3

Table 4.
The Q_n distribution and mean chain lengths (MCL) of C—S—H with various Ca/Si ratios [54].

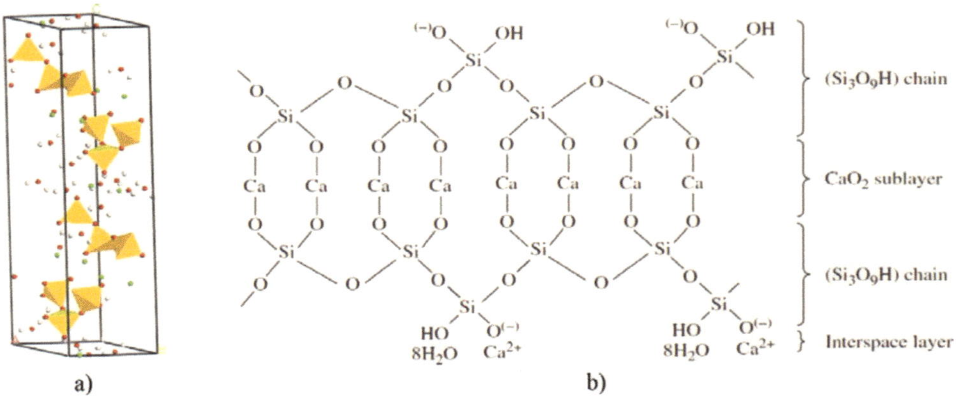

Figure 6.
The structure and atomic connections of 14 Å tobermorite structure [55]. (a) The real cell of 14Å tobermorite by Bonaccorsi [55]. (b) Silicon chain links and atomic connections.

although the chemical aspect CaO should always be CaO, which implies that the oxygen in CaO_2 also includes that of the silicate tetrahedron part). Moreover, silicate chains envelope the Ca—O sheet on both sides, with the characteristics mentioned later. Last but not least, between individual layers, Ca^{2+} and H_2O are filled into the space to balance the charges and determine the layer distance, respectively.

For modeling of the jennite structure [56] (C/S = 1.5), the ordered jennite is triclinic, space group P-1, with the unit cell dimensions of a = 10.576 Å, b = 7.265 Å, c = 10.931 Å, α = 101.3°, β = 96.98°, and γ = 109.65° [57]. The chemical constitutional formula is $Ca_9Si_6O_{18}(OH)_6{\cdot}8H_2O$ and Ca/Si ratio is 1.5. **Figure 7** shows the structure and atomic connections of jennite unit cell.

We can see in **Figure 7** that jennite has the same link of the silicate tetrahedral chains similar to 14 Å tobermorite [55]. The difference is that only half of the oxygen atoms of the Ca—O layer in jennite structure are connected to the silicate tetrahedral chains, while the other half of the oxygen atoms are linked to the hydroxyl groups. Moreover, the bridging tetrahedra of the silicate chains are connected to the Ca—O layer.

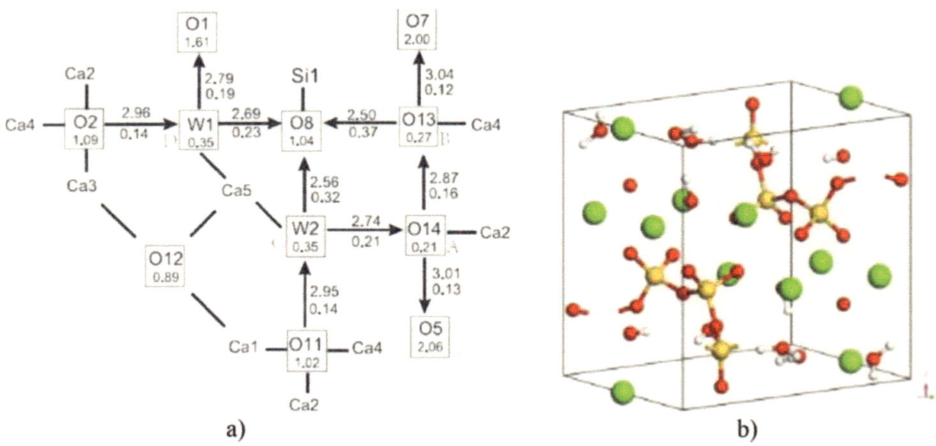

Figure 7.
Modeling structure and atomic connections of jennite crystal structure. (a) The topology structure of jennite crystal [56] and (b) the real structure of jennite crystal [57] (atom types: green, Ca; yellow, Si; red, oxygen; white, hydrogen).

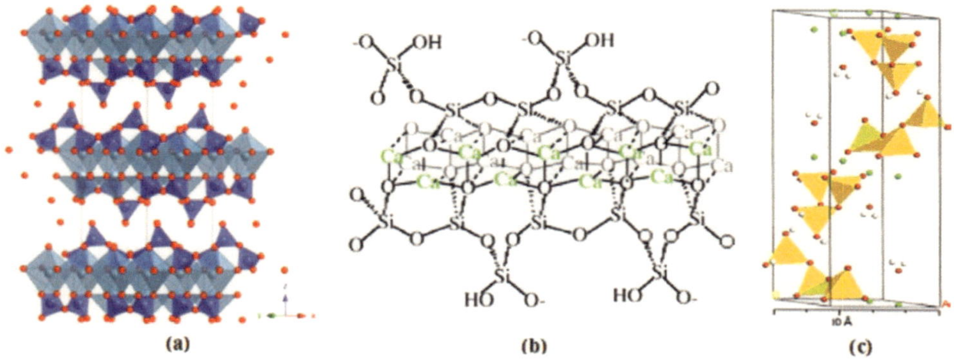

Figure 8.
Modeling of 11 Å tobermorite crystal. (a) 11 Å tobermorite by Hamid [58]; (b) silicon chain links and atomic connections; and (c) the real unit cell.

For modeling of a 11 Å tobermorite: (C/S = 0.67), the morphology of 11 Å tobermorite is monoclinic and the initial lattice is as follows: a = 6.69 Å, b = 7.39 Å, c = 22.779 Å, $\alpha = \beta = 90°$, $\gamma = 123.49°$, its structure is monoclinic with space group P21 [58]. Modeling of 11 Å tobermorite is shown in **Figure 8**.

In **Figure 8(a)**, the 11 Å tobermorite crystal can be summarized as follows: (1) the structure is basically a layered structure with the central part as a Ca—O sheet (with an empirical formula of CaO_2, of which the oxygen in CaO_2 also includes that of the silicate tetrahedron part). (2) Silicate chains envelop the Ca—O sheet on both sides in **Figure 8(b)**. (3) Ca^{2+} and H_2O are filled between individual layers to balance the charges. The unit cell is in **Figure 8(c)**, the infinite layers of calcium polyhedra, which is parallel to (001), with tetrahedral chains of wollastonite-type along b and the composite layers stacked along c and connected through formation of double tetrahedral chains [59].

2.2.3 Radial distribution function (RDF) of C—S—H structures with different Ca/Si ratios

It should be pointed out that the description for C—S—H structures by 11 Å tobermorite structure is not fixed, and the position and the appearance of certain atoms can be changed and adjusted, thus leading to various Ca/Si ratios. There are three different structures with various Ca/Si forms based on Hamid model [60], of which 11 Å tobermorite is denoted as $Ca_8Si_{12}O_{28}(OH)_8{\cdot}4H_2O$ (Ca/Si = 0.67), $Ca_{10}Si_{12}O_{32}(OH)_4{\cdot}4H_2O$ (Ca/Si = 0.83), and $Ca_{12}Si_{12}O_{36}{\cdot}4H_2O$ (Ca/Si = 1.00). The jennite ($Ca_9Si_6O_{18}(OH)_6{\cdot}8H_2O$, Ca/Si = 1.50) is also given [56]. Then a total radial distribution function (RDF) and atomic partial RDF curves of C—S—H structures with different C/S ratios (0.67–1.5) can be calculated, shown in **Figure 9**.

It can be seen in **Figure 9(a)** that, for the system of Ca/Si = 0.66 (or Water/Si = 0.67), four sites of O_8, O_9, O_{14}, O_{18} are protonated caused by the lack of Ca_5 and Ca_6 atoms; For the system of Ca/Si = 0.83 (or Water/Si = 0.50), two sites of O_8, O_9, O_{14}, O_{18} are protonated caused by the lack of Ca_6 atoms. When Ca/Si = 1.0 (or water/Si = 0.33), there are only two water molecules of W_1 and W_2, all sites of O_8, O_9, O_{14}, O_{18} are not protonated because both Ca_5 and Ca_6 atoms are present. In **Figure 9(b)**, there are three kinds of atomic partial RDF curves, of which the chemical bonds include Ca—O, Si—O, Si—Si. The first peak is OH, the second peak is Si—O, the third peak Ca—O, the fourth peak is Si—Si, the respective location is 1.28, 1.56, 2.54, and 3.10 Å, for the interatomic distance between corresponding pairs of atoms.

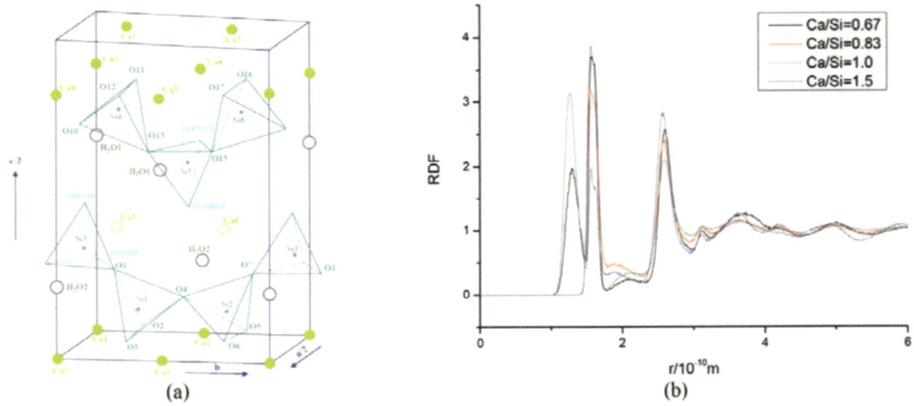

Figure 9.
The total radial distribution function of C—S—H structures with different Ca/Si ratios. (a) A three-dimensional view of tobermorite structure [58] and (b) the total RDF of C—S—H with different Ca/Si ratios.

3. Construction of "glouble" C—S—H by 11 Å tobermorite with various C/S ratios

As the initial configuration is monoclinic, in order to study the deformation of independence in a particular direction, the orthogonal box conversion of the C—S—H structure is needed from the initial monoclinic one, shown in **Figure 10**.

From **Figure 1(a)**), it is noteworthy that the relative atomic position and model size are unchanged in C—S—H model after box transformation in **Figure 1(b)** with the conversion modes of orthogonal box by LAMMPS. Here the conversion is done to obtain the orthogonal structure as an analog initial structure.

3.1 Construction of crystalline 11 Å tobermorite structures (C/S = 0.67)

In order to obtain an amorphous structure in the long-range disorder and short-range order, silicon-oxygen tetrahedra chain is rotated and twisted. Above all, the system was a preequilibrated ensemble of NVT system at 3000 K, and the temperature of the system was gradually lowered to 300 K. After 300 K rebalance, the output is written as an input of the *NVE* heald, then we obtain stable amorphous

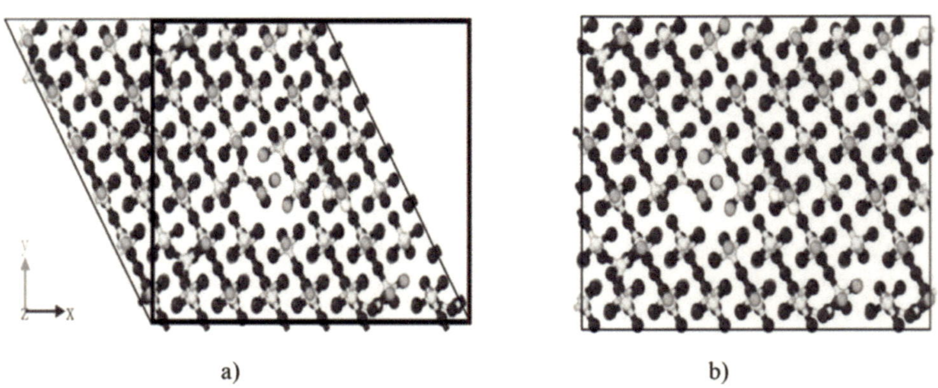

Figure 10.
Schematic conversion diagram of simulation box in x-y plane. (a) Original box of a monoclinic cell and (b) transformed box of a orthogonal cell.

C—S—H. Finally, the simulation results of the C—S—H at 300 K in terms of structural properties and dynamical behavior are analyzed.

Construction of a monolithic C—S—H cell structure corresponding to "globule" C—S—H is in **Figure 11**.

As in **Figure 11**, based on single crystal of 11 Å tobermorite (C/S = 0.67) in **Figure 3(a)** discussed in Chapter 3, two supercells $4a \times 3b \times 1c$ in **Figure 3(b)** and $4a \times 3b \times 2c$ of 11 Å tobermorite model are established first. Then the amorphous model of 5.5 nm^3 is obtained after MD simulation considering an annealing treatment, shown in **Figure 3(c)**. MD simulation of monolithic C—S—H is seen as the "glouble" C—S—H.

The 11 Å tobermorite structure is transformed from crystalline to amorphous by means of an annealing process simulated by molecular dynamics simulation. The modeling procedure is deeply presented in a recent publication of the authors [2], that is why only a brief description is developed here.

An assembly of a supercell with the size of $8a \times 6b \times 2c$ ($5.352 \times 4.434 \times 4.554$ nm^3) of single 11 Å tobermorite cells. These dimensions are chosen to be coherent with the model of Jennings, which describes the C—S—H gel as a packing, more or less dense, of monolithic C—S—H (with a nanogranular aspect of about 5 nm diameter) [49].

Molecular dynamics is a computational simulation method, which aims to study the physical movements of atoms and molecules, by resolving Newton's equations. LAMMPS software is used to perform this study. ClayFF potential is retained to simulate the atomic force field between the various atoms of the supercell.

The various steps to simulate the annealing process and to obtain an amorphous structure are:

- energy minimization (conjugate gradient method) and balance of the structure at 300 K separately under NVT and NVE systems;

- increase of the temperature until 3000 K under NVT system, then balance of the structure at this high temperature until NVT and NVE systems;

- decrease of the temperature from 3000 to 300 K under NVT system;

- balance of the structure at 300 K under NVT and NVE systems.

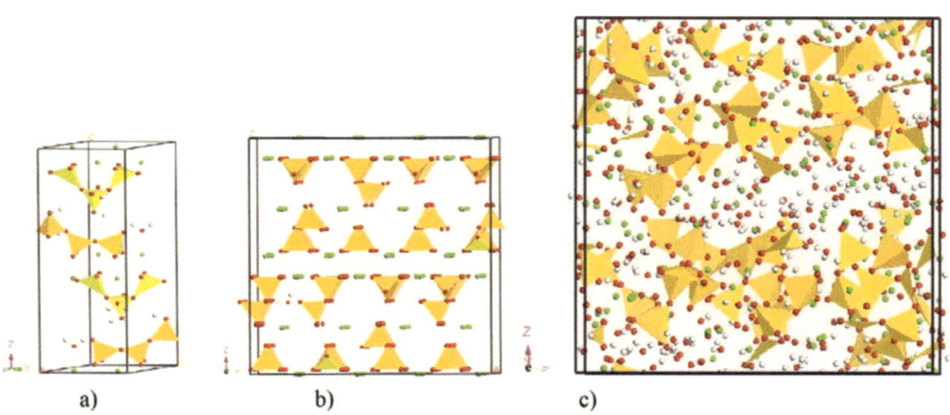

Figure 11.
*Nanoscale modeling of C—S—H with a size of a "globule" by Jennings. (a) 11 Å tobermorite, (b) model of 2.5 nm^3, (c) amorphous model of 5.5 nm^3 (atom types: red-*Ow + *white-*Hw → *water molecule (H_2O); red, O; green, Ca; purple, Ca; yellow, Si).*

After this annealing treatment, the ordered silicon chain becomes disordered and the system is more stable in comparison with the initial structure.

3.2 Construction of amorphous 11 Å tobermorite structure (C/S = 1.67)

Modeling of an amorphous C—S—H structure (C/S = 1.67) with various sizes is in **Figure 12**.

In **Figure 12(a)**, a 3D image of a C—S—H structure with a C/S ratio equal to 1.67 and with a distribution of Q_0 = 13%, Q_1 = 67%, and Q_2 = 20% is presented. Amorphous C—S—H (C/S = 1.67) models in **Figure 12(c)** are constructed by periodic extension in *x*, *y* and *z* directions of the primary cell of $5.352 \times 4.434 \times 4.556$ nm^3 with 8328 atoms presented in **Figure 12(b)**, with an initial density of 2.257 g/cm^3.

Based on the layered supercell of 11 Å tobermorite ($Ca_6[Si_6O_{18}]\cdot 2H_2O$, Ca/Si = 1.0), an amorphous C—S—H(II) (C/S = 1.67) supercell ($4a \times 3b \times 1c$) is constructed, as shown in **Figure 13**.

Figure 13 records the C—S—H model construction procedures, which are as follows: above all, from **Figure 13(a)**, a supercell of $4a \times 3b \times 1c$ 11 Å tobermorite is periodically established by Materials Studio software. Then a cleave surface is built and makes it possible to change the box into an orthogonal type [61], which can be used to better understand mechanical bahaviors of cement paste using MD method by Murray et al. [62]. The simplified formula is $(CaO)_{1.67}(SiO_2)(H_2O)_{1.75}$, which is close to the reference [28]. As presented in **Figure 13(b)** and **Figure 13(c)**, the light silicate tetrahedrals are broken along [010] and [001] and the layered supercell of 11 Å tobermorite without water is taken as the initial configuration. Secondly, **Figure 13(d)** describes a dry C—S—H in which the continuous silicate chains were broken to achieve Q_0, Q_1, and Q_2 percentages of 13, 67, and 20%, respectively, without considering the O—H bonds.

Finally, Grand Canonical Monte Carlo (GCMC) simulation of the water adsorption [63] by LAMMPS using the ClayFF parameters (in **Tables 1–3**) is carried out to obtain the disordered C—S—H structure in **Figure 13(e)** and **(f)**), which is performed on the initially dry C—S—H structure at 300 K. The number of atoms is about 4164, with the initial density of 2.56 g/cm^3. The hydrated phase is presented at various stages of the simulation in **Figure 13(d)–(f)**). At the equilibrium state, the C—S—H reached saturation shown in **Figure 13(f)**), which is also the same with the disordered C—S—H structure shown in **Figure 13(c)**). The final chemical composition and density of the amorphous disordered C—S—H are $(CaO)_{1.67}(SiO_2)$

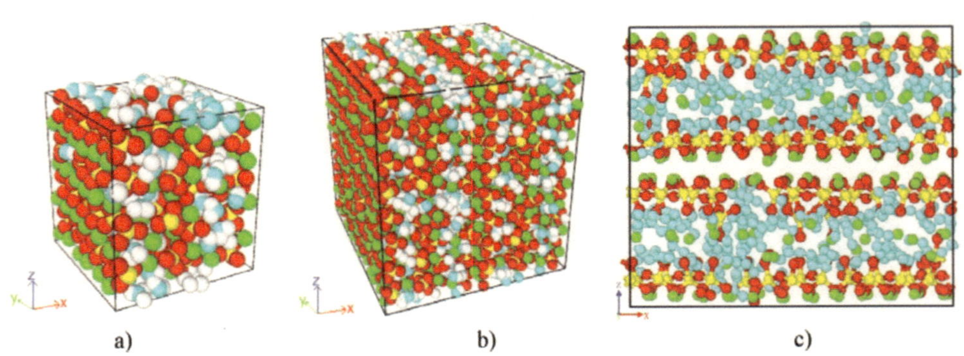

Figure 12.
*Modeling of an amorphous C—S—H structure (C/S = 1.67) with various sizes. (a) A 2.676 × 2.217 × 2.278 nm^3 model, (b) a 5.352 × 4.434 × 4.556 nm^3 model and (c) amorphous 5.5 nm^3 model (atom types: light blue-*Ow *+ white-*Hw *→ water molecule (H_2O); red, O; green, Ca; purple, Ca; yellow, Si).*

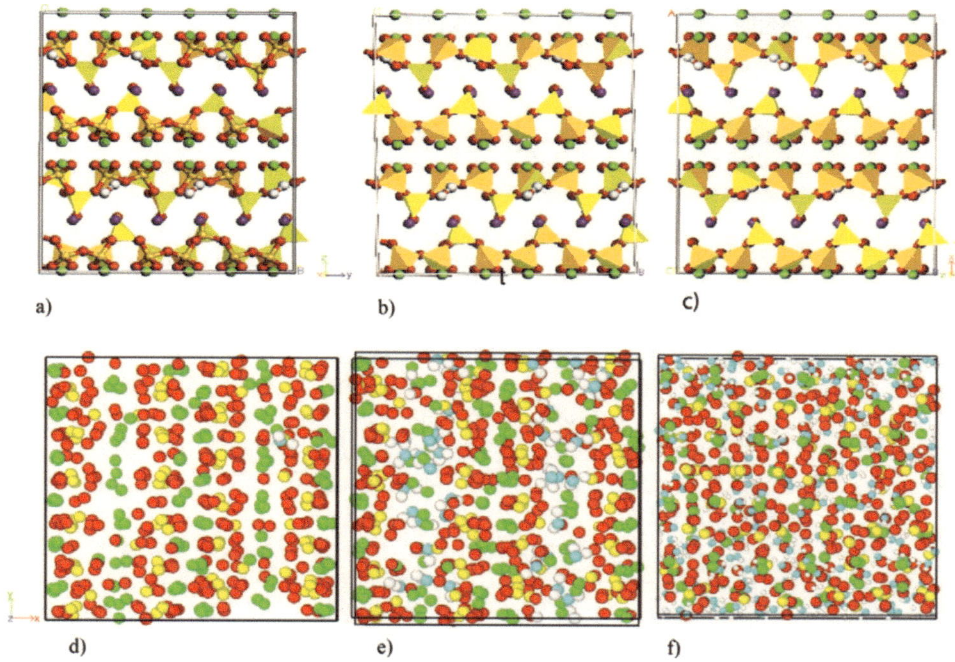

Figure 13.
*Construction of an amorphous C—S—H (C/S = 1.67) supercell of 4a × 3b × 1c 11 Å Tob. (a) Initial 2.676 × 2.217 × 2.278 nm³ model; (b) removal of light silicate tetrahedral along [010]; (c) along [001] (atom types: red-*Ow *+ white-*Hw *→ water molecule (H_2O); red, O; green, Ca; purple, Ca; yellow, Si). (d) GCMC water adsorption (initial); (e) GCMC water adsorption (half); (f) final amorphous C—S—H (atom types: light blue-*Ow *+ white-*Hw *→ water molecule (H_2O); red, O; green, Ca; yellow, Si).*

$(H_2O)_{1.75}$ and 2.257 g/cm³, which are in a good agreement with the results obtained from the neutral scattering tests [22]. As for a C/S ratio of 0.67, this supercell represents the monolithic full-dense C—S—H, in other words, the "globules" by Jennings [49].

4. Simulation and properties of mechanical two kinds of C—S—H (I) (C/S = 0.67)

In order to get the stress-strain relation of C—S—H, the model was subjected to uniaxial tensile loading through gradual elongation with a strain rate of 0.001 ps^{-1}. In the whole simulation process, NPT ensembles are set for the system. Nosé Hoover thermostat temperature control method and the Nosé Hoover barostat pressure method were used for *NPT* calculations. A time step of 1 femtosecond (fs) was chosen since 1 fs is suitable for most purposes. The dynamic time ranges from 100 to approximately 400 picoseconds (ps) depending on the size of the simulation cell, which ensures thermodynamic equilibrium and the convergence of energy with a reasonable computational time. For example, the procedures in the uniaxial tension test in the x direction are as follows. Firstly, the supercells are relaxed at 300 K and coupled to zero external pressure in the x, y, and z dimensions for 50 ps. Then, after the pressures in three directions reach equilibrium, the C—S—H structure is elongated in the x direction. Meanwhile, the pressure in y and z direction should be kept at zero. Pressure evolution in the x direction is taken as the internal stress σ_{xx}. The pressure perpendicular to the tensile direction is set to zero, which allows the normal direction to relax anisotropically without any restriction. The setting, considering the Poisson's ratio, can eliminate the artificial constraint for the deformation. Additionally, in the simulation process, 10,000 configurations are recorded for structural analysis.

4.1 Tensional simulation of the crystaline and amorphous C—S—H(I)

4.1.1 Tensional simulation of the crystalline C—S—H(I)

The crystalline 11 Å tobermorite structure is obtained first and then the tensional deformation under a strain rate of 10^{-3} ps^{-1} is simulated. We assume that, for a tensile strain rate of 10^{-3} ps^{-1}, quasistatic conditions are obtained. The crystaline C—S—H(I) structure ($4a \times 3b \times 1c$) by 11 Å tobermorite is obtained first, and then the tensional deformation under the strain rate 1.0×10^{-3} ps^{-1} is simulated. The tensile deformation process of C—S—H(I) structure with the size of $2.676 \times 2.217 \times 2.278$ nm^3 are calculated and analyzed. Take the *z* direction, for example, the structure evolution of the difference strain under the strain rate of 1×10^{-3} ps^{-1} is in **Figure 14**.

From **Figure 14**, it is indicated that the water molecule has great impact on the mechanical properties of C—S—H(I) during tensile deformation. Moreover, for crystalline C—S—H(I) by 11 Å tobermorite structure ($4a \times 3b \times 1c$) with the size of $2.676 \times 2.217 \times 2.278$ nm^3, it is likely to be broken in the *z* direction, while it is difficult to break in the *x* and *y* directions.

During the stretching process, the aqueous layer is easily broken along the *z* direction, and the stratified phenomenon becomes apparent at both sides. A comparative study of the behaviors of crystalline and amorphous C—S—H(I) is presented in this section. In order to maintain comparative structure, use the same initial structure with the size of $2.676 \times 2.217 \times 2.278$ nm^3.

4.1.2 Tensional simulation of an amorphous C—S—H(I) (C/S = 0.67)

After annealing treatment on Lammps, the ordered silicon chain becomes disordered and the system becomes more stable compared with the initial structure, which is consistent with the real structure of C—S—H(I). Then the stretching process of an amorphous C—S—H (the ratio of Ca/Si is 0.67) at a certain strain rate 1×10^{-3} ps^{-1} is simulated. Take the *z* direction, for example, the nanostructure evolution of the difference strain under the 1×10^{-3} ps^{-1} is shown in **Figure 15**.

From **Figure 15**, it is indicated that the water molecule has a great impact on the mechanical properties of C—S—H(I) during tensile deformation. With an increasing strain, the water molecules in an aqueous layer have a tendency to be pulled off, which may explain the softening behaviors of C—S—H(I) during tensile process. In the structure after annealing, the distribution of water molecules changes a lot, and water molecules become dispersed disorderly. During stretching process, the

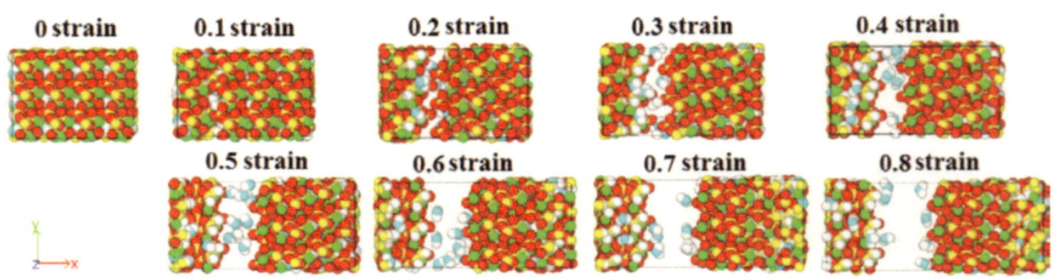

Figure 14.
Results of a crystalline C—S—H(I) structure for various strains under 10^{-3} ps^{-1} *strain rate.*

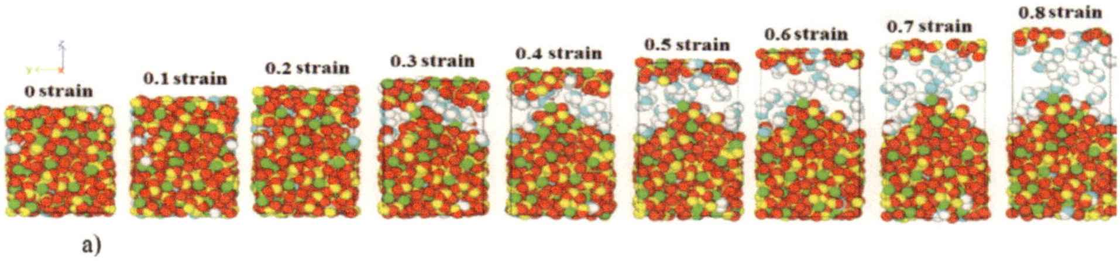

Figure 15.
*Aggregation of water molecules during tensile deformation under different strains. Deformation during a tensional test of a 4a × 3b × 1c supercell model after annealing processes (atom types: light blue-*Ow + *white-*Hw → *water molecule (H_2O); red, O; green, Ca; yellow, Si).*

aqueous layer is only formed in the upper layer, and the deformation resistance is much greater than that in *z* direction in the crystalline structure.

4.2 Elastic moduli results of the crystalline and amorphous C—S—H(I)

4.2.1 Elastic moduli for a crystalline C—S—H(I) (C/S = 0.67) in various directions

The tensional process of an amorphous C—S—H(I) is calculated and analyzed. The corresponding tensile and compressive stress-strain curves are in **Figure 16**.

As is shown in **Figure 16**, for the crystaline C—S—H(I) structure (the supercell 4*a* × 3*b* × 1*c* structure by 11 Å tobermorite) using the ClayFF parameters, Young's modulus in three directions is averaged to about 79.95GPa, which is close to the value of 79.51 GPa obtained by DFT seen in Chapter 3. Similarly, by using BHM potential, the influencing rule of loading direction on strain-stress curves during the elastic linear strain is close to the results by using ClayFF parameters. The elastic moduli in the corresponding *x*, *y*, and *z* directions are 65.61, 82.97, and 85.92 GPa, respectively. Moreover, parameters of ClayFF field seem more stable than those of BHM field (not shown here).

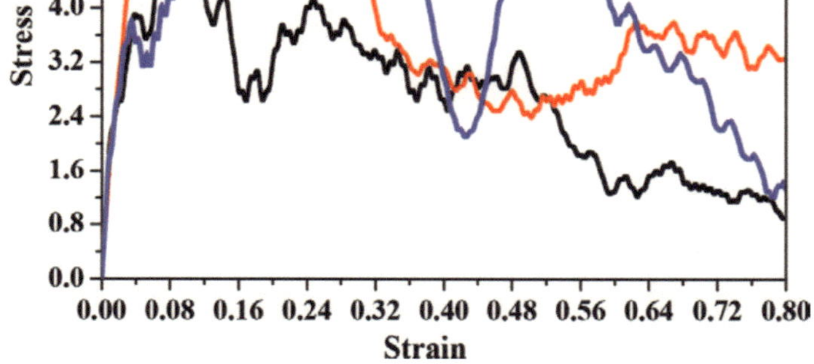

Figure 16.
Results of a crystalline C—S—H(I) structure under strain rate of 10^{-3}.

4.2.2 Elastic moduli for an amorphous C—S—H(I) (C/S = 0.67) in various directions

With the same force field, the tensile test is then simulated in the three directions. The stretch and compression process of an amorphous C—S—H(I) structure ($4a \times 3b \times 1c$ supercell of 11 Å tobermorite, $5.352 \times 4.434 \times 4.556$ nm^3) are calculated and analyzed. The corresponding tensile stress-strain curves are in **Figure 17**.

From **Figure 17**, the three curves are close to each other, showing the isotropy of the supercell as well as its amorphous structure. From these results, by determining the slope at the beginning of these curves (in between 10 and 50% of the maximal stress), the homogeneous Young's modulus may be determined: it is found to be equal to 59.4 GPa. The annealing process leads to a decrease of this elastic property from 79.5 to 59.4 GPa; this result is in agreement with previous results of Vandamme et al. [65]. This value will be used as homogenized Young's modulus to calculate the specific heat of monolithic C—S—H.

In a word, in the disordered structure obtained by annealing treatment, the anisotropy in three directions significantly decreased. It can be inferred that, if the disordered structure becomes much larger, it leads to the isotropic behaviors in a certain degree.

4.2.3 Comparison of crystalline and amorphous C—S—H(I) during tensional process

We can see from **Figures 16** and **Figures 17** that, under the strain rate of 10^{-3} ps^{-1}, peak stress does not change obviously before and after annealing. Result comparison of crystalline and amorphous 11 Å tobermorite supercells during tensional process is shown in **Table 5**.

From **Table 5**, we can see that after annealing process, Young's modulus of C—S—H(I) (C/S = 0.67) structure with a size of $2.676 \times 2.217 \times 2.278$ nm^3 using BHM potential slightly decreased from 78.17 to 68.49 GPa. However, with the ClayFF potential, a decrease of the Young's modulus after annealing process under the strain rate of 10^{-3} has not been pointed out.

Similarly, by using the ClayFF potential, the influencing role of loading direction on strain-stress curves during the elastic linear strain is very close to the results by using BHM potential. After annealing treatment on Lammps, the elastic modulus using the ClayFF field in the corresponding x, y, and z directions are averaged to be 67.89, 84.63, and 70.53 GPa, respectively. Moreover, for C—S—H(I) structure with

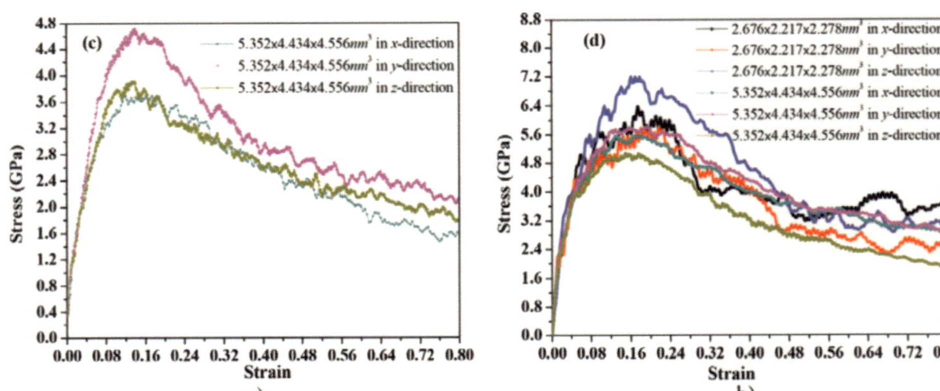

Figure 17.
Results of an amorphous C—S—H(I) structure after annealing by ClayFF. (a) Supercell structure of $5.352 \times 4.434 \times 4.556$ nm^3 and (b) Comparison of C—S—H(I) with different sizes.

Crystal/ structure	Initial size (nm^3)	Cell/box type	Method	E_x	E_y	E_z	$E = (E_x + E_y + E_z)/3$
11Å tobermorite	0.669 × 0.739 × 2.278	Single crystal	DFT	79.512			79.512
11Å tobermorite [60]	0.669 × 0.739 × 2.278	Single crystal	DFT	78.939			78.939
C—S—H(I) (C/S = 0.67)	2.676 × 2.217 × 2.278	Crystalline	ClayFF (MD)	55.82	86.63	97.40	79.95
C—S—H(I) (C/S = 0.67)	2.676 × 2.217 × 2.278	Disordered	ClayFF (MD)	67.89	84.63	78.62	77.05
C—S—H(I) (C/S = 0.67)	5.352 × 4.434 × 4.556	Disordered	ClayFF (MD)	57.85	66.13	54.14	59.37

Table 5.
Elastic modulus comparison of C—S—H(I) structures after optimization.

various sizes at nanoscale in **Table 5**, it seems that the size has a significant influence on the Young's modulus of the amorphous material.

In all, as a kind of C—S—H(I), 11 Å tobermorite supercell (C/S = 0.67) becomes a denser and more stable configuration after annealing treatment, which is in agreement with the previously proposed quantitative "colloid" model of C—S—H gel, resulting in an improved understanding of the microstructural changes associated with drying and heat curing [64].

5. Simulation and mechanical properties of an amorphous C—S—H(II) (C/S = 1.67)

Before simulation, it must be also noted that the total energy obtained by this method (called lattice energy for crystalline systems) has no physical sense for most force fields. After repeated tests, ClayFF force field is used to complete the MD simulation.

5.1 Tensional simulation of amorphous C—S—H(II) structures

The stretching of the C—S—H structure of 5.352 × 4.434 × 4.556 nm^3 at the strain rate of 10^{-3} ps^{-1} under various directions is shown in **Figure 18**.

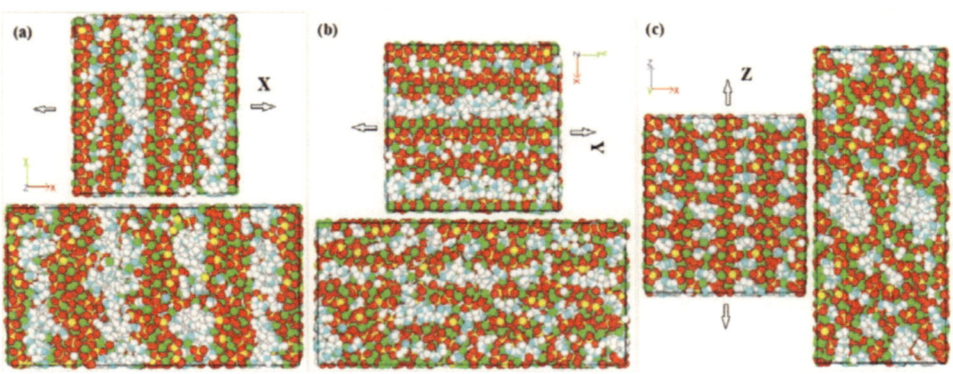

Figure 18.
The stretching C—S—H structure at the strain rate of 10^{-3} ps^{-1} under various directions.

From **Figure 18**, the aggregation of water molecules changes in different directions during the stretching process. The layers of water molecules in the *x* direction are extended and the width of the water molecules layer becomes larger, while the water molecule layer in the *y* direction is narrowed. The water molecules gathered toward the center of the structure in the *z* direction; the different states of aggregation form of water molecules in various directions cause the anisotropic properties of C—S—H structure with size of $5.352 \times 4.434 \times 4.556$ nm^3.

5.2 Elastic moduli of the amorphous C—S—H(II) determination

In order to better understand the influence of size on the performance of C—S—H(II) structures (C/S = 1.67), the stretching process is simulated to obtain stress-strain curves and then to find the influencing rules of these parameters on the properties of C—S—H(II) structures (C/S = 1.67).

5.2.1 Effect of loading direction on strain-stress curves of size $5.352 \times 4.434 \times 4.556$ nm^3

Figure 19 shows the tensional stress-strain curves in *x*, *y*, and *z* directions of an amorphous C—S—H structure of size $5.352 \times 4.434 \times 4.556$ nm^3 under the strain rate of 10^{-3} ps^{-1}.

Figure 19(a) shows the strain-stress curves of a supercell structure with size of $5.352 \times 4.434 \times 4.556$ nm^3 under various directions. From **Figure 19(b)**, we can see that the maximum stress of tensional stress-strain curve with the size of $2.676 \times 2.217 \times 2.278$ nm^3 is higher than the other size of $5.352 \times 4.434 \times 4.556$ nm^3. With the increase of size, the peak stress becomes lower. It can also be seen that the presence of the maximum stress in stress-strain curves greatly varies due to the scale effect between atomic interactions. Young's modulus with the size of $5.352 \times 4.434 \times 4.556$ nm^3 is lower than the other size of $2.676 \times 2.217 \times 2.278$ nm^3, which may be due to the size effect at nanoscale. However, there is little change of stress-strain curve slope in the initial elastic stage. Moreover, the averaged modulus in three directions is 60.95 GPa under strain rate of 10^{-3} at 300 K, which is close to the other results of the C—S—H structure with the value of 63.50 [25] and 69.01 GPa [55]. A supercell of $5.352 \times 4.434 \times 4.556$ nm^3 as a monolithic full-dense C—S—H (globules) appears to be sufficient to determine Young's modulus, and it can be used to further estimate Young's moduli of LD and HD C—S—H, which is the aim of this chapter.

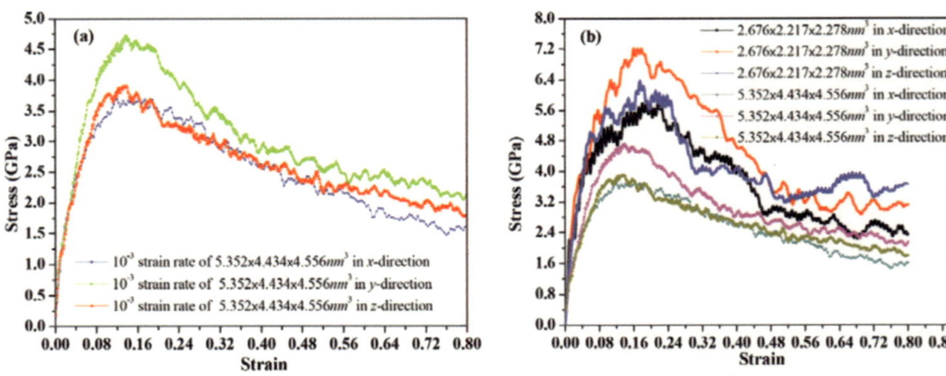

Figure 19.
Comparison of supercell structures under various directions by MD simulation. (a) Supercell structure of $5.352 \times 4.434 \times 4.556$ nm^3 and (b) comparison of two supercell structures.

Crystal/structure	Initial size (nm^3)	Box type	Method	E_x	E_y	E_z	$(E_x + E_y + E_z)/3$
C—S—H(I) (C/S = 0.67)	2.676 × 2.217 × 2.278	Crystalline	MD	55.82	86.63	97.40	79.95
C—S—H(I) (C/S = 0.67)	5.352 × 4.434 × 4.556	Disordered	MD	57.85	66.13	54.14	59.37
C—S—H(II) (C/S = 1.67)	5.352 × 4.434 × 4.556	Disordered	MD	55.82	58.89	68.13	60.95

Table 6.
Elastic modulus comparison of C—S—H structures constructed by 11 Å tobermorite.

Elastic modulus comparison of C—S—H constructed by 11 Å tobermorite is in **Table 6**.

From **Table 6**, we can see that the deformation direction has a little influence on the mechanical properties of the amorphous C—S—H(II) (C/S = 1.67) under the strain rate of 10^{-3} ps^{-1}. Meanwhile, for the structure of C—S—H(II) (C/S = 1.67) with size of 5.352 × 4.434 × 4.556 nm^3, the corresponding averaged elastic modulus in y and z directions is higher than that in x direction by the stress-strain curve slope within about 0.002–0.055 strain. Moreover, there is only a slight difference between elastic moduli of C—S—H(I) and C—S—H(II). For all results of the C—S—H(I) structures in three directions mentioned in **Table 5**, the averaged value is very close to each other by using ClayFF parameters. Conversely, the size has a significant influence on the Young's modulus of the amorphous C—S—H(II) structure.

5.2.2 Assessment of the elastic properties of amorphous C—S—H structures by MD

As the solid phase described by Jennings is named by "globules," which was found to have a characteristic size of about 5.6 nm [49]. Here we can use the elastic modulus of the CSH(II) structure with a size of 5.352 × 4.434 × 4.556 nm^3 to approximate the modulus of "globules" so as to investigate the LD C—S—H and HD C—S—H.

A self-consistent scheme may be used [65]:

$$\frac{K}{k_s} = \frac{4G(1-\phi_s)}{4G + 3k_s\phi_s} \tag{16}$$

$$\frac{G}{g_s} = \frac{1}{2} - \frac{5}{4}\phi_s - \frac{3}{16}r_s(3-\phi_s) + \frac{1}{16}\sqrt{64 - 320\phi_s + 400\phi_s^2 + r_s\left(144 - 168\phi_s - 120\phi_s^2\right) + r_s^2\left(81 - 54\phi_s + 9\phi_s^2\right)} \tag{17}$$

where $r_s = \frac{k_s}{g_s} = \frac{2(1+\mu_s)}{3(1-2\mu_s)}$, g_s is shear modulus of the monolithic full-dense C—S—H (here $g_s = E_s/[2(1+\mu_s)]$ with $E_s = 60.95GPa$ and $\mu_s = 0.25$ for C—S—H(II) found in this work). k_s is bulk modulus of the monolithic full-dense C—S—H (here $k_s = E_s/[3(1-2\mu_s)]$ with $E_s = 60.95GPa$ and $\mu_s = 0.25$ for C—S—H(II) structure).

The Mori-Tanaka scheme may also be used [65]:

$$K = \frac{\sum_r f_r k_r \left(1 + \alpha_s\left(\frac{k_r}{k_s} - 1\right)\right)^{-1}}{\sum_r f_r \left(1 + \alpha_s\left(\frac{k_r}{k_s} - 1\right)\right)^{-1}} \tag{18}$$

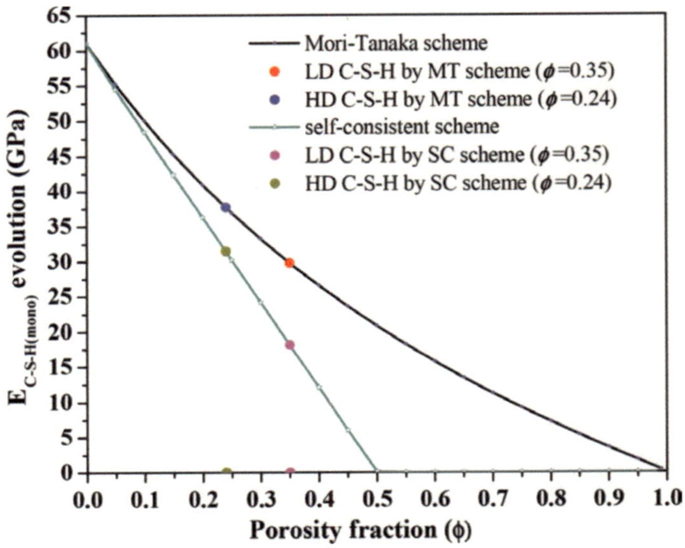

Figure 20.
Young's modulus evolution of C—S—H(II) (C/S = 1.67) according to the porosity.

$$G = \frac{\sum_r f_r k_r \left(1+\beta_s\left(\frac{g_r}{g_s}-1\right)\right)^{-1}}{\sum_r f_r \left(1+\beta_s\left(\frac{g_r}{g_s}-1\right)\right)^{-1}} \tag{19}$$

$$\alpha_s = \frac{3k_s}{3k_s + 4g_s} \tag{20}$$

$$\beta_s = \frac{6(k_s + 2g_s)}{5(3k_s + 4g_s)} \tag{21}$$

where α_s and β_s are coefficients, f_r is the volume fraction, k is the bulk modulus, g is the shear modulus. The subscripts "o" stands for the monolithic C—S—H.

Figure 20 presents the evolution of the Young's modulus according to the porosity.

As is shown in **Figure 20**, Young's modulus of LD C—S—H and HD C—S—H can be estimated by a self-consistent scheme and the Mori-Tanaka scheme. For example, with the self-consistent scheme in Eqs. (16) and (17), and assuming a porosity of, respectively, 0.35 and 0.24, the Young's moduli of LD C—S—H and HD C—S—H are found to be equal to 18.11 and 31.45 GPa, which is in agreement with the results of Constantinides (2006). Similarly, by using the Mori-Tanaka scheme in Eqs. (18)–(21), the values of Young's moduli of LD C—S—H and HD C—S—H are, respectively, 29.78 and 37.71 GPa.

6. Conclusions

In this chapter, the stretch process of a represented amorphous full-dense C—S—H(I) and C—S—H(II) with the Ca/Si ratios of 0.67 and 1.67 is simulated by LAMMPS program after setting parameters of related potential functions. Some conclusions can be drawn.

1. For two typical crystalline and amorphous C—S—H(I) structures (formula: $Ca_4Si_6O_{14}(OH)_4{\cdot}2H_2O$), the stretch process at a certain strain rate is calculated, and the corresponding tensile stress-strain curves are analyzed. The results are as follows:

- Before annealing process, the crystalline supercell ($4a \times 3b \times 1c$) of C—S—H(I) at nanoscale has the averaged elastic modulus value of 79.95 GPa by MD simulation, which is similar to the homogenized Young's modulus value of 79.512 GPa by elastic constants of single 11Å tobermorite model based on DFT.

- Through annealing process of the C—S—H(I) with a size of $2.676 \times 2.217 \times 2.278\ nm^3$, using ClayFF field parameters, the difference of Young's modulus in three directions becomes smaller during the tensional deformation, with an averaged value of 77.05 GPa. For the C—S—H(I) structure with a size of about 5 nm, using MD simulation, the averaged Young's moduli are separately about 59.37 GPa.

- By comparison of both the crystalline and amorphous C—S—H(I) structures, it is found that the water molecule in both structures has a great influence on tensile strength.

- After annealing simulation, the annealed structure becomes more uniform in three directions. Young's modulus is very slightly reduced; however, we think that with a larger supercell corresponding to globule C—S—H, such a decrease would be obtained, indicating that the annealed supercell with disordered silicon chain is more softened and more consistent in x, y and z directions.

2. A typical C—S—H(II) structure (formula: $(CaO)_{1.67}(SiO_2)(H_2O)_{1.75}$) is simulated by using the ClayFF field. Above all, a layered supercell of 11Å tobermorite of Hamid model without water is firstly considered as the initial configuration. Then, SiO_2 groups in silica tetrahedra have been removed, guided by the NMR results to achieve realistic percentages of Q_0, Q_1, and Q_2. Finally, Grand Canonical Monte Carlo simulation of water absorption on the dry C—S—H structure at 300 K is performed to obtain a disordered C—S—H structure.

    - The water molecule in the structure has a great influence on tensile strength.

    - It is difficult to find a RVE to assess the tensile strength and the postpeak part of the tensile curve. However, Young's modulus of C—S—H(II) structure with a size of about 5 nm has been identified to be between 55 and 68 GPa (the averaged modulus in three directions is 60.95 GPa under strain rate of $10^{-3}\ ps^{-1}$ at 300 K), which is consistent with previous literature results. The Young's modulus in x, y, and z directions is rather the same, which suggests that the model is indeed amorphous.

    - The effect of size on the response of the amorphous C—S—H structure has been also investigated; by the variation of stress-strain curve due to the atomic interactions, the presence of scale effect is observed (this means that, with the increase of size, the peak stress becomes lower). Moreover, Young's modulus with the size of $5.352 \times 4.434 \times 4.556\ nm^3$ is lower than the other size of $2.676 \times 2.217 \times 2.278\ nm^3$.

    - Water layer is an important factor, which affects the mechanical properties, showing the weakest loading resistance perpendicular to the aqueous layer.

Overall, we mainly considered the C/S influencing factor, which has an effect on the stiffness and strength of C—S—H gel. The presence of impurities in C—S—H gel (i.e., sulfate or aluminate), porosity, or a specified density should be also considered in future work.

3. The values we have found for the Young's moduli of amorphous C—S—H(I) and C—S—H(II) are those of the monolithic C—S—H (globules). At the upper scale, we can deduce the Young's moduli of LD C—S—H and HD C—S—H assuming a porosity for these two kinds of C—S—H. This study is limited to the C—S—H(II), the kind of C—S—H found in the hardened cement pastes experimentally tested and simulation results in Chapter 6. We can see that there is a significant difference between the two homogenization methods, as previously pointed out by Constantinides.

Acknowledgements

The author acknowledges the financial support provided by the China Scholarship Council and the startup foundation of Xi'an Shiyou University. Thanks Qiufeng Wang for her proofreading, and very sincere thankness to prof. Fabrice Bernard and prof. Siham Kamali-Bernard for the their enthusiastic and responsible guideness

Author details

Jia Fu[1,2]

1 Xi'an Shiyou University, Xi'an, China

2 INSA de Rennes, Rennes, France

*Address all correspondence to: fujia@xsyu.edu.cn

References

[1] Fu J, Bernard F, Kamali-Bernard S. First-principles calculations of typical anisotropic cubic and hexagonal structures and homogenized moduli estimation based on the Y-parameter. Application to CaO, MgO, CH and calcite $CaCO_3$. Journal of Physics and Chemistry of Solids. 2017;**101**:274-289

[2] Fu J, Bernard F, Kamali-Bernard S. Assessment of the elastic properties of amorphous calcium silicates hydrates (I) and (II) structures by molecular dynamics simulation. Molecular Simulation. 2018;**44**(4):285-299

[3] Fu J, Kamali-Bernard S, Bernard F, Cornen M. Comparison of mechanical properties of C—S—H and portlandite between nano-indentation experiments and a modelling approach using various simulation techniques. Composite Part B: Engineering. 2018; **151**:127-138

[4] Fu J, Bernard F, Kamali-Bernard S. Multiscale modeling and mechanical properties of zigzag CNT and triple-layer graphene sheet based on atomic finite element method. Journal of Nano Research. 2015;**33**:92-105

[5] Sekkal W, Zaoui A, Benzerzour M, Abriak NE. Role of porosity on the stiffness and stability of (001) surface of the nanogranular C-S–H gel. Cement and Concrete Research. 2016;**87**:45-52

[6] Bernard F, Kamali-Bernard S. Numerical study of ITZ contribution on mechanical behavior and diffusivity of mortars. Computational Materials Science. 2015;**102**:250-257

[7] Keinde D, Kamali-Bernard S, Bernard F. Effect of the interfacial transition zone and the nature of the matrixaggregate interface on the overall elastic and inelastic behaviour of concrete under compression: A 3D numerical study. European Journal of Environmental and Civil Engineering. 2014;**18**(10):1167-1176

[8] Kamali-Bernard S, Keinde D, Bernard F. Effect of aggregate type on the concrete matrix/aggregates interface and its influence on the overall mechanical behavior. A numerical study. Key Engineering Materials. 2014; **617**:14-17

[9] Kamali-Bernard S, Bernard F. Effect of tensile cracking on diffusivity of mortar: 3D numerical modelling. Computational Materials Science. 2009; **47**:178-185

[10] Bernard F, Kamali-Bernard S, Prince W. 3D multi-scale modelling of mechanical behaviour of sound and leached mortar. Cement and Concrete Research. 2008;**38**:449-458

[11] Curtin WA, Miller RE. Atomistic/continuum coupling in computational materials science. Modeling and Simulation in Materials Science and Engineering. 2003;**11**:33-68

[12] Jia F, Huai S. An ab initio force field for predicting hydrogen storage in IRMOF materials. Journal of Physical Chemistry C. 2009;**113**(52):21815-21824

[13] Lu G, Kaxiras E. Overview of multiscale simulations of materials. In: Rieth M, Schommers W, editors. Handbook of Theoretical and Computational Nanothechnology. Vol. X. 2005. pp. 1-33

[14] Ye G. Experimental study and numerical simulation of the development of the microstructure and permeability of cementitious materials [doctoral dissertation]. TU Delft, Delft University of Technology; 2003

[15] Gilman JJ. Chemistry and Physics of Mechanical Hardness. Hoboken: John Wiley & Sons; 2009

[16] Al-Rub RKA, Voyiadjis GZ. Analytical and experimental determination of the material intrinsic length scale of strain gradient plasticity theory from micro-and nano-indentation experiments. International Journal of Plasticity. 2004;**20**:1139-1182

[17] Tho KK, Swaddiwudhipong S, et al. Numerical simulation of indentation with size effect. Materials Science and Engineering: A. 2006;**421**(1–2):268-275

[18] Raki L, Beaudoin JJ, Alizadeh R, Makar JM, Sato T. Cement and concrete nanoscience and nanotechnology. Materials. 2010;**3**:918-942

[19] Kwan S, LaRosa Thompson J, Grutzeck MW. Structures and phase relations of aluminum-substituted calcium silicate hydrate. Journal of the American Ceramic Society. 1996;**79**(4): 967-971

[20] Taylor HFW, Howison JW. Relationship between calcium silicates and clay minerals. Clay Mineral Bulletin. 1956;**31**:98-111

[21] Taylor HFW. Cement Chemistry. 2nd ed. London: Thomas Telford; 2007

[22] Jennings HM. A model for the microstructure of calcium silicate hydrate in cement paste. Cement and Concrete Research. 2000;**30**(1):101-116

[23] Bernard F, Fu J, Kamali-Bernard S. Multiscale modeling approach to determine the specific heat of cementitious materials. European Journal of Environmental and Civil Engineering. 2018;**23**:1-17

[24] Hou D et al. Calcium silicate hydrate from dry to saturated state: Structure, dynamics and mechanical properties. Acta Materialia. 2014;**67**:81-94

[25] Pellenq RJ-M, Lequeux N, Van Damme H. Engineering the bonding scheme in C—S—H: The iono-covalent framework. Cement and Concrete Research. 2008;**38**(2):159-174

[26] Kalinichev AG, Wang J, Kirkpatrick RJ. Molecular dynamics modeling of the structure, dynamics and energetics of mineral–water interfaces: Application to cement materials. Cement and Concrete Research. 2007;**37**(3):337-347

[27] Cygan RT, Liang JJ, Kalinichev AG. Molecular models of hydroxide, oxyhydroxide, and clay phases and the development of a general force field. Journal of Physical Chemistry B. 2004; **108**(4):1255-1266

[28] Pellenq RJ-M, Kushima A, Shahsavari R, et al. A realistic molecular model of cement hydrates. Proceedings of the National Academy of Sciences. 2009;**106**(38):16102-16107

[29] Xin H, Lin W, Fu J, Li W, Wang Z. Temperature effects on tensile and compressive mechanical behaviors of C—S—H structure via atomic simulation. Journal of Nanometerials. 2017;**8476258**:1-6

[30] Shahsavari R, Pellenq RJ-M, Ulm FJ. Empirical force fields for complex hydrated calcio-silicate layered material. Physical Chemistry Chemical Physics. 2011;**13**(3):1002-1011

[31] Ji Q, Pellenq RJ-M, Vlieta KJV. Comparison of computational water models for simulation of calcium–silicate–hydrate. Computational Materials Science. 2012;**53**(1):234-240

[32] Berendsen HJC, van der Spoel D, van Drunen R. GROMACS: A message-passing parallel molecular dynamics implementation. Computer Physics Communications. 1995;**91**(1):43-56

[33] Zaoui A. Insight into elastic behavior of calcium silicate hydrated oxide (C–S–H) under pressure and composition effect. Cement and Concrete Research. 2012;**42**(2):306-312

[34] Zhang W, Wang H, Ye J. Structure and its variation of calcium silicate hydrates. Journal of the Chinese Ceramic Society. 2005;**33**(1):63-68. (In Chinese)

[35] Dai W, Shui Z, Duan P. Study on the structural model of calcium silicate hydrate based on computer simulation. International Conference on Energy and Environmental Science Singapore: Energy Procedia. 2011;**47**:1-5

[36] Al-Ostaz A, Wu W, Cheng AHD, et al. A molecular dynamics and microporomechanics study on the mechanical properties of major constituents of hydrated cement. Composites Part B: Engineering. 2010; **41**(7):543-549

[37] Constantinides G, Ulm F-J. The effect of two types of C—S—H on the elasticity of cement-based materials: Results from nanoindentation and micromechanical modeling. Cement and Concrete Research. 2003;**34**(1):67-80

[38] Gao Y, De Schutter G, Ye G, et al. The ITZ microstructure, thickness and porosity in blended cementitious composite: Effects of curing age, water to binder ratio and aggregate content. Composites Part B: Engineering. 2014; **60**:1-13

[39] Abidi S, Joliff Y, Favotto C. Impact of perlite, vermiculite and cement on the Young modulus of a plaster composite material: Experimental, analytical and numerical approaches. Composites Part B: Engineering. 2016;**92**:28-36

[40] Keinde D. Etude du béton à l'échelle mesoscopique: Simulation numérique et tests de micro-indentation [doctoral dissertation]. INSA de Rennes; 2014. (In French)

[41] Arnaud Charles Albert M. Characterization of porosity & CSH in cement pastes by 1H NMR [Thèse de doctorat]. École Polytechnique Fédérale de Lausanne; 2014

[42] Constantinides G, Ulm FJ. The effect of two types of CSH on the elasticity of cement-based materials: Results from nanoindentation and micromechanical modeling. Cement and Concrete Research. 2004;**34**(1):67-80

[43] Rapaport DC. The Art of Molecular Dynamics Simulation. 2nd ed. Cambridge, New York: Cambridge University Press; 2004

[44] Wei-Zhong L, Cong C, Jian Y. Molecular dynamics simulation of self diffusion coefficient and its relation with temperature using simple Lennard-Jones potential. Heat Transfer—Asian Research. 2008;**37**(2):86-93

[45] Kai L, Zhonghe S, Wei D. A molecular dynamics study on the structure characteristic of calcium silicate hydrate (CSH) gels. In: Computer Science for Environmental Engineering and EcoInformatics. Berlin Heidelberg: Springer; 2011. pp. 33-39

[46] Heyes DM. The Liquid State: Application of Molecular Dynamics. New York: John Wiley and Sons; 1998

[47] Dobson JF. Harmonic-potential theorem: Implications for approximate many-body theories. Physical Review Letters. 1994;**73**(16):2244

[48] Plassard C, Lesniewska E, Pochard I, Nonat A. Investigation of the surface structure and elastic properties of calcium silicate hydrates at the nanoscale. Ultramicroscopy. 2004; **100**(3):331-338

[49] Jennings HM. A model for the microstructure of calcium silicate hydrate in cement paste. Cement and Concrete Research. 2000;**30**(1):101-116

[50] Masoero E, Del Gado E, Pellenq RJM, Yip S, Ulm FJ. Nano-scale mechanics of colloidal C–S–H gels. Soft Matter. 2014;**10**(3):491-499

[51] Richardson IG, Groves GW. The composition and structure of C—S—H in hardened slag cement pastes. In: Proceedings of the 10th ICCC. Vol. II. Sweden: Gothenburg; 1997

[52] Grutzeck MW, LaRosa-Thompson J, Kwan S. Characteristics of CSH gels. In: Proceedings of the 10th International Congress on the Chemistry of Cement, Gothenburg, Sweden. Göteborg, Sweden: Amarkai AB and Congrex Göteborg AB; 1997

[53] Xiandong C, Kirkpatrick RJ. ^{29}Si MAS NMR study of the structure of calcium silicate hydrate. Advanced Cement Based Materials. 1996;**3**(3): 144-156

[54] Tajuelo Rodriguez E, Richardson I G, Black L, et al. Composition, silicate anion structure and morphology of calcium silicate hydrates (CSH) synthesised by silica-lime reaction and by controlled hydration of tricalcium silicate (C3S). Advances in Applied Ceramics. 2015;**114**(7):362-371

[55] Bonaccorsi E, Merlino S, Kampf AR. The crystal structure of tobermorite 14 Å (Plombierite), a C–S–H phase. Journal of the American Ceramic Society. 2005; **88**:505-512

[56] Bonaccorsi E, Merlino S, Taylor HFW. The crystal structure of jennite, $Ca_9Si_6O_{18}(OH)_6 \cdot 8H_2O$. Cement and Concrete Research. 2004;**34**(9):1481-1488

[57] Hajilar S, Shafei B. Molecular Dynamics Simulation of Elastic Properties of Ordered CSH Gel: Case Study of Tobermorite and Jennite. Supplementary Volume of Nanotechnology in Construction. New York: Springer; 2015

[58] Hamid SA. The crystal structure of the 11Å natural tobermorite $Ca_{2.25}[Si_3O_{7.5}(OH)_{1.5}] \cdot 1H_2O$. Zeitschrift für Kristallographie-Crystalline Materials. 1981;**154**(1–4):189-198

[59] Merlino S, Bonaccorsi E, et al. The real structure of tobermorite 11Å normal and anomalous forms, OD character and polytypic modifications. European Journal of Mineralogy. 2001;**13**(3): 577-590

[60] Shahsavari R, Buehler MJ, Pellenq RJM, et al. First-principles study of elastic constants and interlayer interactions of complex hydrated oxides: Case study of tobermorite and jennite. Journal of the American Ceramic Society. 2009;**92**(10): 2323-2330

[61] Janakiram Subramani V, Murray S, Panneer Selvam R, Hall KD. Atomic structure of calcium silicate hydrates using molecular mechanics. In: 88th Annual Meeting of the Transportation Research Board. Washington, D.C.: National Research Council; 2009

[62] Murray S, Subramani V, Selvam R, Hall K. Molecular dynamics to understand the mechanical behavior of cement paste. Transportation Research Record: Journal of the Transportation Research Board. 2010;**2142**:75-82

[63] Puibasset J, Pellenq R. Water adsorption on hydrophilic mesoporous and plane silica substrates: A grand canonical Monte Carlo simulation study. The Journal of Chemical Physics. 2003; **118**(12):5613-5622

[64] Jennings HM, Thomas JJ, Gevrenov JS, Constantinides G, Ulm F-J. A multi-technique inve stigation of the nanoporosity of cement paste. Cement and Concrete Research. 2007;**37**:329-336

[65] Vandamme, M. The nanogranular origin of concrete creep: a nanoindentation investigation of microstructure and fundamental properties of calcium-silicate-hydrates. [PhD Thesis]. Massachusetts Institute of Technology; 2008

Chapter 6

Nanoindentation Investigation of Typical C—S—H Structures: FEM Simulation

Jia Fu

Abstract

This work focuses on elastic modulus of two main constituents of cement-based materials: portlandite (CH) and calcium silicate hydrate (C—S—H). The large-scale models using QC methods are considered trustworthy and on the ABAQUS software, where, on behalf of the purely elastic scene, the unloading curves of the three (CH, LD and HD CSH, CSH) models are simulated. Based on the elastic modulus of the monolithic C—S—H structure by MD simulations, the assessment results on LD C—S—H and HD C—S—H after homogenization are very close to nanoindentation simulation data. The findings are as follows: (1) the homogenized elastic properties of polycrystals can be obtained by elastic constants of single crystal (using DFT and RVH estimation) and thus can be used to explain the relationship between structure and mechanical properties of CH from nano- to microscale. (2) It is found that the results of comparison of the simulated and experimental unloading curves of CH and LD/CH C—S—H are essentially coincident and have a very good agreement, showing the feasibility and the rationality of the simulation methods above. Results enable to provide useful parameters for composite cement system modeling and a method to calculate elastic modulus of other similar structures.

Keywords: calcium silicate hydrate, elastic modulus, nanoindentation, FEM

1. Introduction

Multiscale simulations of atomistic/continuum coupling in computational materials science have become a hot spot for recent years, and the different scales will be considered to reflect the atomic interaction [1]. Multiscale modeling of structural concrete performance is presented as a systematic knowledge base of coupled cementitious composites and structural mechanics [2]. At the scale of micromechanical investigations, the interaction of the various phases is taken into account. In this way, for the greater length scale, a unique constitutive behavior is extracted that typically cannot be captured fully by standard closed-form continuum models [3]. An illustrative way for denoting these different length scales is the micro-, meso-, and macroscale [4]. Among these, the level of components or structural parts is called macroscale. At present, there are two major directions in multiscale modeling [5]: extending the atomistic length scales (spatial multiscale methods) and extending the atomistic time scales (temporal multiscale methods). For the length scale extension, atomistic-continuum coupling may be applied:

atomistic method can be used in the region of localized deformation where resolution down to atomic scale is desired (fine scale domain), and continuum method (e.g., FEM) everywhere else (coarse scale domain).

In theory, the modulus estimation of deformation involves different types of processes, including elastic, plastic, and fracture theories, and its theoretical mechanism is very complex [6]. Therefore, the prediction of hardness and mechanical modulus of deformation has a lot of difficulties in theory. So, the experimental measurements to calculate elastic modulus become the easiest way, mainly a variety of indentation experiments [7]. That is why, the nanoindentation experiment has been widely used to characterize mineral and metal materials [8].

Nanoindentation technology has become an important means to measure material properties and is widely used in the fields of medical devices, microelectro mechanical systems (MEMS), bioengineering, civil engineering, aircraft, and film material research. Indentation technology is the expansion of conventional hardness test to microscale (no more than submillimeter) [9], the principle of it is to get parameters of microindentation hardness and elasticity modulus, etc., according to loading/unloading curves between load and displacement [10, 11]. The disadvantage of the classical plasticity theory is that it has not considered size effect [12, 13]. Models commonly used during micro- and nanoindentation process mainly include elastic recovery model, Kick model [14], proportioned sample model (PSR), Hays-Kendall model, and Taylor dislocation model [15, 16]. In general, the elastic moduli of covalent crystals are intrinsic and equivalent to the sum of resistance of each bond per unit area to indenter. In this chapter, analytical method is used to solve the nanoindentation simulation with indentation depth during nanoindentation process and fitting methods are adopted to establish the finite-element constitutive model.

Using backscattered images and point analysis, outer and inner C—S—H can be distinguished. **Figure 1** shows SEM BSE micrograph of different phases in ordinary Portland cement (OPC) paste at 7 days of hydration [17] and point's position of the paste with w/c = 0.40 after 28 days of hydration for the EDX-SEM analysis [17].

From **Figure 1(a, b)**, it is indicated that the outer C—S—H has a brighter gray level, while inner C—S—H is mainly localized within the original cement grains. As is shown in **Figure 1(c, d)**, there are mainly four regions of hydrated products, such as: voids region (the cumulative hydration voids), C—S—H region (including low density and high density), CH phase region, and RC phases region (the residual cement in a cement paste).

Elastic moduli and hardness can be used to understand the extent, to which a solid resists both elastic and plastic deformation. C—S—H gel has high surface areas and shows excellent adhesive characteristics. Nanoindentation can be used to understand differences in hydration products that result from water availability for clinker during hydration. Ye et al. [18] have simulated the microstructure of cementitious materials by using the cement hydration model-HYMOSTRUC. Ulm and coworkers [19, 20] have provided the experimental techniques and analysis tools to provide reliable, mechanical technical tools to identify the relative volume fractions and spatial distributions of cement hydration products in cement paste samples produced with conventional cements. Besides, they have illustrated that there are three types of C—S—H (LD C—S—H, HD C—S—H, C—S—H/CH nanocomposite) in hardened cement paste. The average nanoscale material responses (e.g., indentation modulus and indentation hardness) for each of these three C—S—H phases are virtually the same for all Portland cement pastes tested to date [21, 22].

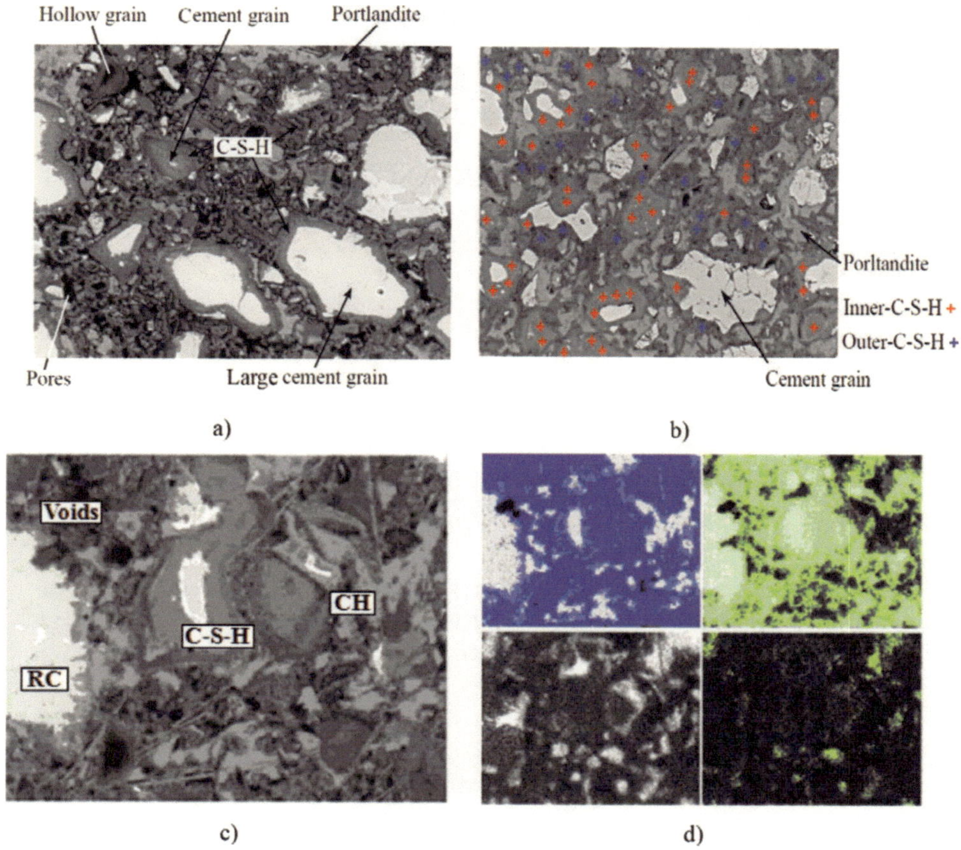

Figure 1.
BSE micrograph of typical phases and the C—S—H points in cement paste. (a) SEM BSE micrograph of an OPC paste [17]. (b) BSE micrograph map of the inner and outer C—S—H points [17]. (c) SEM BSE micrograph of the OPC paste [17]. (d) Regional distribution of chemical elements [17].

Nanoindentation method may provide a wide variety of mechanical characteristics without substrate effects to obtain information on the mechanical properties of hard, soft, brittle, and ductile materials at the nanometer scale. The techniques have been widely used in the characterization of small-scale mechanical behavior of materials, which provides an ideal approach for measuring phases at nanoscale upon a substrate [23, 24]. Although there are many reports on nanoindentation, however, these studies are almost based on experimental analysis and its numerical calculation at nanoscale has not yet been revealed. Multiscale modeling and simulation of atomistic/continuum coupling in computational CH, LD C—S—H, and HD C—S—H structures with the size of nanometer unit during nonlinear indentation is considered and simulated to compare the results of Keinde [25] and Constantinides [26]. In this chapter, nanoindentation simulations are carried out to determine elastic moduli of several phases in cement pastes, compared with results of nanoindentation experiments. As the volume fraction of main components of CH and C—S—H in hydrated Portland cement paste separately reach more than 14 and 40%, our main task is to predict the Young's modulus via multiscale simulation and nanoindentation experiment.

2. Background of theory, principle, and measurement of nanoindentation

2.1 Numerical description of nanoindentation procedure

During indentation, the load and displacement were continuously monitored by a three-plate capacitive force/displacement transducer. Nanoindentation hardness is defined as the maximum indentation load P_{max} divided by the contact area A_c:

$$H = \frac{P_{\max}}{A_c(h_c)} \tag{1}$$

The indenter shape function A (nm^2) can be seen as a function of the indentation depth h_c:

$$A = f(h_c) \tag{2}$$

Here, area function selected depends on the function of Oliver and Pharr proposed and modified considering loading system flexibility and curvature radius of the indenter tip correction. So, the projected contact area calculation is as:

$$A_p = C_0 \cdot {h_c}^2 + C_1 \cdot h_c + C_2 \cdot {h_c}^{1/3} + C_3 \cdot {h_c}^{1/7} + \ldots\ldots \tag{3}$$

where C_1, C_2 ... C_n are coefficients describing geometrical shape of indenter, of which high major is used to describe deviation degree of relative ideal shape at indenter tip.

The unloading stiffness S is then established by differentiating P at the maximum depth of penetration. S is the initial unloading contact stiffness, which is represented by the slope of the initial portion of the unloading curve and described as:

$$S = \left(\frac{dP}{dh}\right)_{h=h_{\max}} \tag{4}$$

The contact area as a function of contact depth h_c is calibrated by indenting on a standard cement specimen. The contact depth h_c can be estimated from the load-displacement data using Sneddon's equation [27] and assuming that pile-up is negligible:

$$h_c = h_{\max} - \varepsilon \frac{P_{\max}}{S} \tag{5}$$

where ε is a constant that depends on the indenter geometry; ε = 0.75 for the Berkovich indenter, ε = 0.72 for the conical indenter, and ε = 1 for the flat punch. P_{max} is the maximum load and S is the slope of unloading curve.

The indentation depth $h_{\max}$ at any time during loading can be described as follows:

$$h_{\max} = h_c + h_s = h_c + \varepsilon \cdot \frac{P_{\max}}{S} \tag{6}$$

where h_s is the displacement (nm) over the initial surface nearby the section between indenter and tested sample. The reduced modulus of material is defined as:

$$E_r = \frac{\sqrt{\pi}}{2\beta}\frac{S}{\sqrt{A}} \tag{7}$$

which can be further expressed as:

$$\frac{1}{E_r} = \frac{1-\nu^2}{E} + \frac{1-\nu_i^2}{E_i} \tag{8}$$

where h_c is the contact depth, E_s and v_s are the elastic modulus and Poisson's ratio of the specimen, and E_i and μ_i are the elastic modulus and Poisson's ratio of the indenter. For a diamond indenter, $E_i = 1140\ GPa$ and $\mu_i = 0.07$. The unloading curve defined by Oliver and Pharr is described by a power law relation:

$$P = A\left(h - h_f\right)^m \tag{9}$$

where A and m are fitting parameters empirically determined.

It can also be noted that Young's modulus E and Poisson coefficient μ are linked in Eq. (8), and without any assumption on the value of one of these characteristics, the other one cannot be determined by the nanoindentation test.

2.2 Principle of the finite-element analytical method

Guillonneau [28] has discussed the process of nanoindentation in detail both the measurement and the numerical simulation, where FEM method is used to know the deformation distribution in the contact area.

Figure 2 is a schematic diagram of rigid conical indenter and its geometrically necessary dislocations underneath. Assuming that the indenter is rigid, microplastic deformation caused by geometric necessary dislocations underneath illustrated in **Figure 2** can be expressed as follows:

$$\tan\theta = h/a = b/s, s = b \cdot a/h \tag{10}$$

where θ is the angle between indenter surface and sample surface, a is indenter radius contacted, h is indentation depth, b stands for Burgers vector, and s represents distance within a slipped step of indentation.

Huang et al. [29] have developed the model of Nix and Gao [30] and proposed a new theoretical model based on the maximum geometric necessary dislocation density allowed. Assuming that λ is the total length of dislocation loops, integration from r to $r + dr$ within dislocation loops is carried out in the whole contacted radius a of indenter and λ can be described as follows:

$$\lambda = \int_0^a \frac{2\pi r}{s}dr = \int_0^a \frac{h}{b \cdot a} \cdot 2\pi r dr = \frac{h}{b \cdot a} \cdot \pi r^2 \Big|_0^a = \frac{h \cdot a}{b}\pi \tag{11}$$

Considering size effect, the intrinsic material length is introduced. Assuming that flow stress is under control of microscale dislocation motion and compliances with Taylor hardening relation, the equivalent stress in microscale can be expressed as: $\overline{\sigma} = \sigma_y\sqrt{f^2(\overline{\varepsilon}) + l\eta}$. The model of Nix and Gao et al. [30] on the basis of continuum mechanics can be described as follows:

$$\sigma = K\sqrt{\varepsilon_p{}^{2n} + l\eta} = K\sqrt{\varepsilon_p{}^{2n} + l \cdot (\rho_G \cdot b)} \tag{12}$$

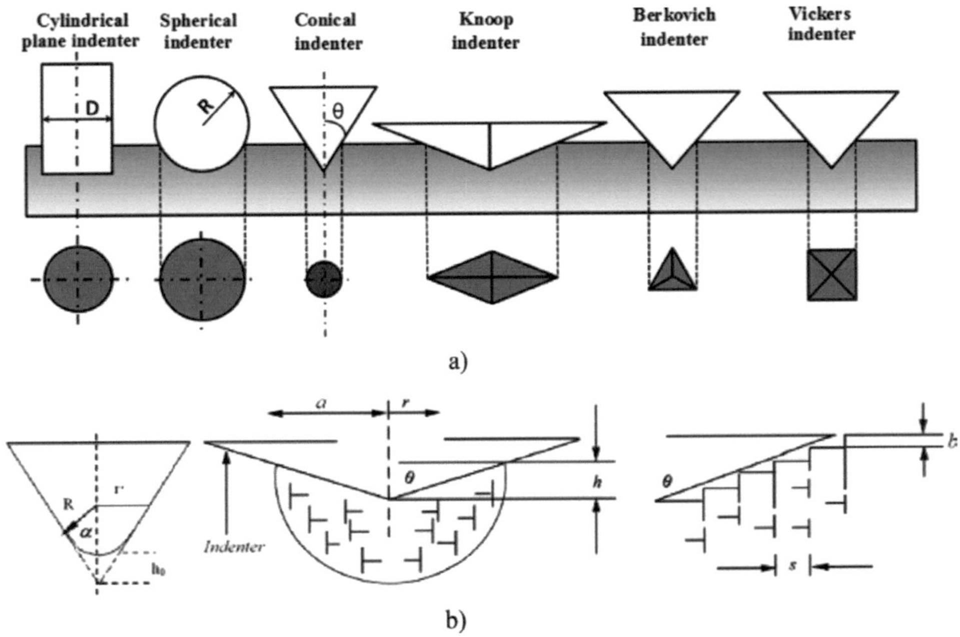

Figure 2.
Schematic diagram of rigid conical indenter and its dislocations region underneath [15]. (a) Different schemas and different types of indentation points [28]. (b) Schematic diagram of rigid conical indenter and its dislocations region underneath [15].

where σ represents stress, K is strengthening coefficient, n is strain hardening exponent, b is Burgers vector, ρ_G is geometric dislocation density, ε_p is the peak of effective strain, η is equivalent strain gradient, expressed as $\eta = \rho_G \cdot b$, and l is intrinsic material length (microns) and can be expressed as: $l = \left(M\alpha\mu/\sigma_{ref}\right)^2 \cdot b$.

Assuming that all dislocation loops are within the volume V of hemisphere with its radius a, the geometrically necessary dislocation density caused by the ideal conical rigid indenter with the sample being pressed can be described as follows:

$$\rho_G = \frac{\lambda}{V} = \frac{(a \cdot h \cdot \pi)/b}{2a^3 \cdot \pi/3} = \frac{3h}{2a^2 \cdot b} = \frac{3}{2bh}\tan^2\theta \tag{13}$$

where h is the indenter depth, and θ is the angle between indenter surface and sample surface with its value 19.7°.

Assuming that conical indenter is rigid and the contact surface with indenter is under friction-free state, the schematic diagrams of load state and displacement state brought by conical indenter with material pressed can be expressed in **Figure 3**.

The indentation schematic diagrams using Berkovich indenter are shown in **Figure 3(a)**. Take the spherical indenter for example, as is illustrated in **Figure 3(b)**, the displacement boundary on contact sphere and the contact boundary of interfacial nodes at the top of indenter must keep the sliding condition at spherical part. Moreover, the displacement boundary on conical surface and the contact boundary of interfacial nodes at conical part of indenter must keep the sliding condition too. So, the displacement boundary of node i at spherical part and the incremental displacement of node j at conical part under the incremental calculation step n during microindentation process can be described as follows:

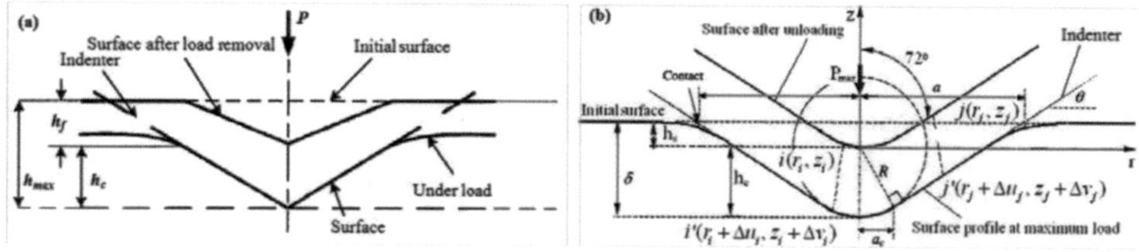

Figure 3.
Schematic diagrams of displacement and load during microindentation. (a) Conical indenter. (b) Spherical indenter.

$$-\frac{r_i}{\sqrt{a^2-{r_i}^2}}\cdot\Delta u_i+\Delta v_i=-\delta_n \tag{14}$$

$$-\Delta u_j/tg\theta+\Delta v_j=-\delta_n \tag{15}$$

where r and z are cylindrical coordinates, (r_i, z_i) stands for coordinate of node i; Δu_i and Δv_i are incremental node displacements in r and z directions, a is the hemispherical radius of indenter, δ_n is incremental displacement caused by indenter with material being pressed, θ is half-cone angle, and the value is 70.3° for Berkovich indenter.

Assuming that conical indenter is fully rigid and there is no friction between indenter and material [31], as the contact surface stress of sample pressed by indenter is at normal direction, so the load of node can be adopted approximately at normal direction. The nodal force at node N can be divided into F_{Nr} and F_{Nz} at directions of r and z, so the incremental loads at directions of r and z can be separately expressed as ΔF_{Nr} and ΔF_{Nz}. The incremental loads of spherical contact node i and conical contact node j can be separately described as follows:

$$\Delta F_{ir}\cdot\cos\alpha+\Delta F_{iz}\cdot\cos\alpha=0 \tag{16}$$

$$\Delta F_{jr}+\Delta F_{jz}\cdot tg\theta=0 \tag{17}$$

The modified incremental load ΔF_e must be added in the direction of radius to keep the deformation load along the direction of normal at contact surface required, the expression of ΔF_e can be described as follows:

$$\Delta F_e=-F_{ir}-F_{iz}\cdot tg\alpha=0 \tag{18}$$

where exists the equation: $\alpha=arctg\left(r_i/\sqrt{a^2-r_i^2}\right)$.

For the regular resolution of contacted fine-grid section, the incremental finite element equation can be described as:

$$[K]\{\Delta d\}=\{\Delta F\} \tag{19}$$

where *[K]* is total stiffness matrix, Δd and ΔF, respectively, represent incremental displacement and load. Insert load boundary conditions and displacement boundary conditions of corresponding contact surface nodes into Eq. (19) and Eqs. (14), (16), and (18) are inserted for the ball contact surface node i while the proportionally conical surface contact node j with Eqs. (15) and (17) inserted, so the microindentation hardness obtained ultimately can be resolved by the simultaneous equations described as:

$$\begin{bmatrix} k_{2i-1,1}+k_{2i,1}\cdot tg\alpha & \dots & k_{2i-1,2i-1}+k_{2i,2i-1}\cdot tg\alpha & k_{2i-1,2i}+k_{2i,2i}\cdot tg\alpha \\ 0 & \dots & -r_i/\sqrt{a^2-r_i^2} & 1 \\ k_{2j-1,1}+k_{2j,1}/tg\theta & \dots & k_{2j-1,2j-1}+k_{2j,2j-1}/tg\theta & k_{2j-1,2j}+k_{2j,2j}/tg\theta \\ 0 & \dots & -1/tg\theta & 1 \end{bmatrix} \tag{20}$$

$$\times \begin{bmatrix} \Delta u_i \\ \Delta v_i \\ \Delta u_j \\ \Delta v_j \end{bmatrix} = \begin{bmatrix} -F_{ir}-F_{iz}\cdot tg\alpha \\ -\delta_n \\ 0 \\ -\delta_n \end{bmatrix}$$

Solving equations of the contact nodes in Eq. (14) as well as noncontact nodes treated by conventional computing method, displacement distribution during unloading from P_{max} can be finally obtained.

Assuming that the material follow Mises flow law and the Tabor factor [32] is equal to 3, the microindentation hardness of material can be converted by equivalent stress described by shear strength with the dislocation density [32] according to the model of Nix and Gao [30] described the Taylor hardening relationship by dislocation density, so the equation of microindentation hardness can be described as follows:

$$H = 3\overline{\sigma} = 3\sqrt{3}\overline{\tau} = 3\sqrt{3\alpha\cdot\mu\cdot b\cdot\sqrt{\rho_S+\rho_G}} \tag{21}$$

where ρ_S is the statistics stored dislocation density, μ is the shear modulus, and α is an empirical constant, whose magnitude order value is one.

As experimental microindentation hardness values are related with maximum load P_{max} (mN) and indenter shape function A (nm^2), the relationship between relative microindentation hardness values and indentation depth considering the intrinsic material length l without considering the effect of indentation elastic recovery can be expressed as follows:

$$\frac{H}{H_0} = \sqrt{1+\frac{h^*}{h}} = \sqrt{1+k\cdot\frac{h^*}{d}} \tag{22}$$

where H is indentation hardness (GPa), H_0 is indentation hardness without considering strain gradient effect, h is indentation depth, d is the reciprocal of indentation diagonal length, k is identified as $4\tan\theta/\sqrt{2}$, and A is area function.

The expressions of A, H, H_0, and h were separately described as:

$$H_0 = 3\sqrt{3}\alpha\cdot\mu\cdot b\sqrt{\rho_S} \tag{23}$$

$$h^* = \frac{81}{2}b\cdot\alpha^2\cdot\tan^2\alpha\cdot\left(\frac{\mu}{H_0}\right) \tag{24}$$

$$H = P_{\max}/A \tag{25}$$

$$A = P_{\max}/\sum_{n=0}^{S} C_n\cdot h_c^{1/(2^n-1)} \tag{26}$$

In case of analytical extraction of elastic properties through nanoindentation data, it is possible to assess roughly the yield stress as being directly correlated with hardness value according to Tabor [33], and given by:

$$\sigma_y \approx H/C \tag{27}$$

where H is indentation hardness (GPa), and C is the contact creep modulus (GPa); the coefficient can change in the range of 2.6–3.0.

3. Nanoindentation modeling and simulation at nano-/microscale

3.1 Nanoindentation modeling using the QC method

3.1.1 Representative volume element (RVE)

In general, elastic properties (Young's modulus, Poisson ratio, etc.), tensile strength, and postfailure tensile stress-displacement are thus the unique data that are transferred from cement paste to mortar [34]. Besides, the elastic displacements are found in every pixel, and the average strain and stress is computed and averaged over the entire microstructure to give the effective elastic properties. Input data of these solid phases at the nanoscale are needed. Different scales in structural modeling of concrete using HYMOSTRUC model of multiscale structures by UT Delft [35] and MuMoCC platform [35, 36] are developed. The representative volume element primarily uses numerical approximations of the effective properties of composite. Square or cubic RVEs are used for most numerical approximations because of the ease of numerically solving boundary value problems with these geometries. The difficulties involved in generating statistical information about particle distributions and concentrations lead to difficulties in the rigorous determination of RVE sizes. Sab [37] has shown that if an RVE exists for a random composite material, the homogenized properties of the material can be calculated by the simulation of one single realization of the medium. The "ergodic" hypothesis assumes that the ensemble average is equal to the volume average. The ensemble average is the mean of a large number of realizations of the microstructure. The volume average, on the other hand, is the average as the RVE volume becomes infinitely large compared to the volume of a particle.

The MuMoCC platform operates in two main steps. The first one consists of generating realistic 3-D representative volume elements of concrete and their three different scales (cement paste, mortar, and concrete). This hierarchical modeling is based on the assumption of scale separation which is for the moment necessary. Indeed, because of the wide range of feature size in concrete, it is impossible to represent concurrently all these structural features in a single mesomodel and then to represent the complete system as one. The NIST 3-D model, CEMHYD3D, is then used for this purpose. The MuMoCC platform enables one to couple this well-known model to the FE software ABAQUS. A set of algorithms has been developed in order to convert the voxelized images obtained in this first step by CEMHYD3D to FE meshes compatible with ABAQUS. **Figure 4** shows a 2-D cross section of the generated submesostructure (mortar) and the modeled microstructure (hydrated cement paste).

From **Figure 4(a)**, we can see that, at the mesoscale of mortar, large aggregates with interfacial transition zone (ITZ) embedded in a matrix of mortar can typically be observed. From **Figure 4(b)**, at microscale, the hydrated cement paste corresponding to the considered mortar is modeled using NIST's CEMHYD3D code. As also seen before, this paste is a heterogeneous and complex porous medium in which the main solid phases are calcium silicate hydrate (C—S—H), Portlandite (calcium hydroxides) and hydrated aluminates, and sulfoaluminate phases. The solid phases are in chemical equilibrium with an interstitial solution that partially or

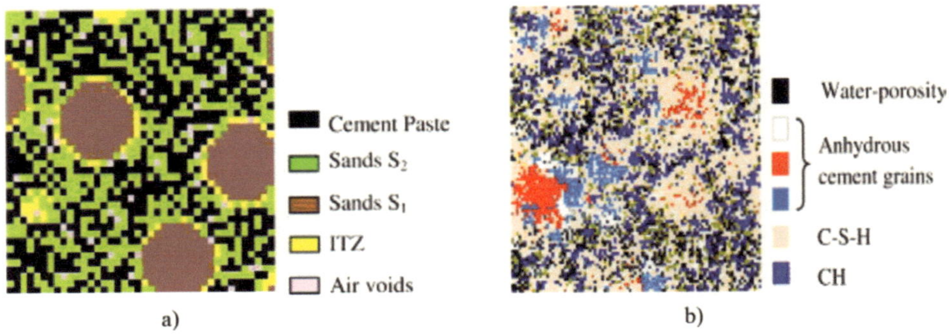

Figure 4.
Structural modeling of the meso- and microstructure models [38, 39]. (a) Mesostructure model of mortar (125 × 125 μm); (b) microstructure model of the cement paste (1 × 1 μm).

totally fills the porosity. The second step of the method consists of imposing various boundary conditions on the meshed RVE. The failure in mortar is supposed to be the consequence of crack initiation and propagation at highly localized regions under large tensile stress concentration. Failure in compression is supposed to be the consequence of tensile stresses resulting from restrained deformations due to the heterogeneous mesostructure of mortar. Thus, the tensile behavior of the hardened cement paste is needed as input data to bridge the scales.

Thus, the quasicontinuum (QC) method is a mixed continuum and atomistic approach for simulating the mechanical response of polycrystalline materials. For nanoindentation modeling using the QC method, the representative volume element (RVE) is used for nanoindentation simulation.

3.1.2 The 2-D/3-D modeling and axisymmetric finite element simulations

The main eight steps of nanoindentation simulation using ABAQUS finite element analysis software are mainly divided as:

1. ABAQUS CAE component module: A solid model composed by indenter and the sample is established first. The sample is established as a 2-D/3-D axisymmetric deformable part, and the indenter is thought as rigid since the hardness and elastic modulus of the indenter material is far greater than the sample material. For Berkovich indenter model, the half-cone model with a degree of 70.32° is used, thus has the same relation between area and penetration of nanoindentation experiment. In order to eliminate the influence of outside boundary conditions on result of the indentation region, the sample length of each side is set 40 times the indentation depth.

2. Characteristics module: Material properties (elastic constants, plastic, etc.) are defined.

3. Assembly module: Model instances of indenter and sample are defined and assembled.

4. Grid module: The samples are meshed and defined by certain element types. In order to better simulate the deformation behavior of the indentation affected zone, setting with a bias grid, unit type is quadrilateral reduced integration unit with the type of CPE4R element.

5. Analysis step module: The unloading step, time step, and analysis step are defined.

6. Interaction module: Related properties can be defined. In this chapter, indenter surface is set as the main surface, and the surface of sample is subjected to the main surface. The friction coefficient between contact surfaces may be defined.

7. Load module: The bottom surface of the sample is fixed, the left axis is of symmetry, and boundary conditions are applied to the contacted indenter point.

8. Analysis job module: The job is established and then submitted to analysis solver.

The nanoindentation constitutive model with embedded programs was used to simulate nanoindentation process by finite element analysis. The meshing diagram is shown in **Figure 5**.

Based on the steps mentioned above, 2-D/3-D axisymmetric finite element simulations are performed to investigate the effect of friction on the indentation response of phases, e.g., C—S—H [38]. The indenter was modeled as a rigid cone with half apex angle of 70.32°. Details of the model geometry are shown in **Figure 5**. This conical angle ensures the same contact depth-projected area of contact relation as in Vickers and Berkovich pyramidal indenters.

3.2 Typical *P-h* curve analysis of the elastic modulus

The penetration depth is measured in nanometers continuously during the loading process in nanoindentation simulation. Apart from the scale movement, the distinguishing feature of most nanoindentation testing is the indirect measurement of the contact area (real area where contact is made indenter). In conventional indentation test, the contact area is calculated by measuring the size of the residual footprint on the surface of the sample after unloading. In the nanoindentation test, the contact surface is determined so indirectly from the geometry of the indenter and the depth of penetration; meanwhile, the force and the displacement of the indentor in the material are continuously recorded and shown in a force-depth curve, as in **Figure 6**.

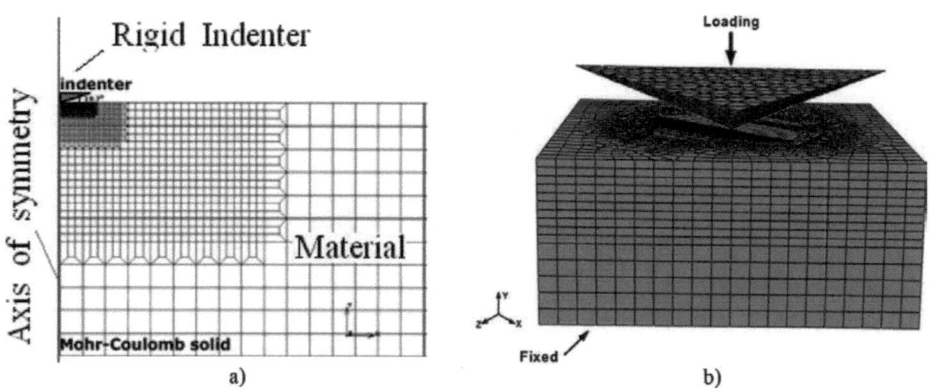

Figure 5.
Indentation simulation on 2-D/3-D models of typical cement paste phases. (a) 2-D geometrical details and finite element mesh. (b) 3-D geometrical details and finite element mesh.

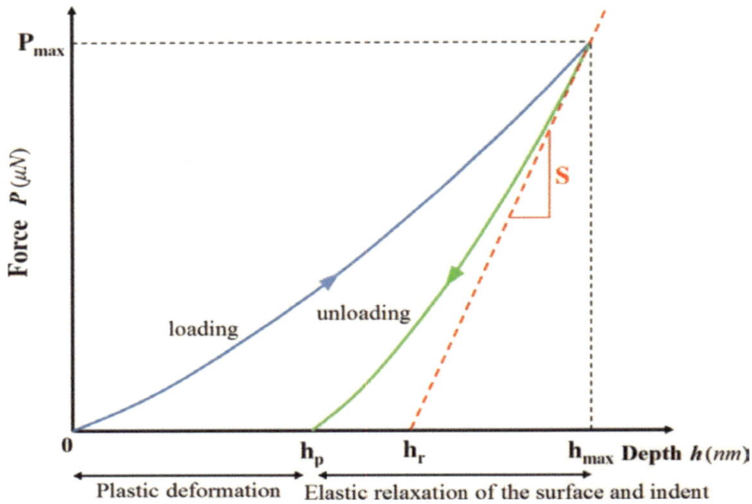

Figure 6.
Typical load-depth curves in nanoindentation experiment.

As is shown in **Figure 6**, S is the initial unloading contact stiffness, h_p is the residual depth during plastic deformation, h_c is the contact depth, P_{max} is the maximum of loading stress, and h_{max} is the maximum of indentation depth. The typical indentation curve in **Figure 6** represents a loading-unloading cycle, not superimposed when the test leaves a permanent imprint on the surface. The first part is the loading curve corresponding to the penetration of the indentor. This ends when the indenter reached its maximum load P_{max} or the maximum penetration h_{max}. The second part represents the discharge curve. It corresponds to the withdrawal of the indenter. The analysis of a cycle gives a lot of information by reading the indentation curve total penetration depth h_{max} and depth of the residual depth h_p. The area enclosed by load curve, unload curve, and horizontal axis represents plastic deformation work, while the area enclosed by unload curve, horizontal axis, and the vertical ordinate of maximum load represents elastic deformation work.

During the nanoindentation simulation, the movement can apply forces from micronewtons up to several millinewtons and measurement depths down to a few nanometers. To achieve such good resolution at nanometer length, the model should be generally placed in a soundproof enclosure on an antivibration table to eliminate the negative influence of vibrations and noise so as to obtain the correspondingly accurate load-depth curves which we need.

4. Nanoindentation technique for CH and C—S—H structures by FEM

Nanoindentation method is a relatively simple and effective technique to evaluate the mechanical properties of material, and it can not only obtain the relevant performance parameters of the material but also reflect the mechanism of elastic-plastic behaviors and reveal relationship between microstructure and macroscopic mechanical properties. However, it has a complex problem of contact, which is affected by many factors, such as nanoindentation test, the surface roughness, the substrate effect, the grain boundary effects, indenter geometry, lattice anisotropy, size effect, etc. Even if the nanoindentation process is tested in the same device and under the same test conditions, the result cannot be guaranteed repeatability. That is why, the investigation of nanoindentation process research using numerical simulation method is required.

4.1 Models of nanoindentation on 2-D/3-D models

The nanoindentation constitutive model with embedded programs was used to simulate nanoindentation process by FE analysis. At the mesoscale, subdomains are regarded with locally varying loads, and interaction of the constituents is accounted for at the microscale [39]. For simplicity purposes, the Berkovich indenter is commonly modeled as a conical indenter with a semiapex angle of 70.32°, where the conical cross-sectional area closest to the actual situation [40]. According to literatures [41, 42], 3-D models can be equivalent to an axisymmetric 2-D model and the results of two models are consistent with each other. Thus, the nano-indentation 3-D model is treated as an axial symmetrical 2-D model, shown in **Figure 7**.

From **Figure 7(a)**, the 3-D model has a size of 15.0 × 15.0 × 7.5 μm, which has the sufficient space so as to compare the experimental results. The CPE4R elements with higher accuracy and axisymmetric model are adopted to simulate nanoindentation process. Berkovich indenter has an angle of 70.32°. To simplify the calculation, Berkovich indenter is assumed to be rigid and the deformation is subjected to the Mises yield criterion and isotropic hardening criterion. The size of microindentation model regarded as the semiinfinite solid in **Figure 7(b)** is 12.0 × 12.0 μm. This continuum space is discretized using four-noded axisymmetric, isoparametric element (CPE4R element). A mesh sensitivity analysis was performed to ensure that the simulation results were insensitive to the mesh size in the indenter region. To save the computational time, the transitional gird division method is adopted. The element size was continuously refined as approaching the indenter contact region for greater accuracy, a refined mesh in **Figure 7(c)** was near the tip area (contact zone is 3.0 × 3.0 μm, the number of grid is 300 × 300) in order to minimize the computing resource and to maintain the reasonable accuracy. The minimum length of elements within the final domain is about 10 nm.

4.2 Parameters of typical structures in simulation

Correspondingly, the indentation is simulated by ABAQUS software at the same depth from the simulation. The applied displacement force is added. Mechanical parameters of typical phases in simulation are found in the previous chapters and listed in **Tables 1** and **2**.

In the following simulations, as in the reference of Asroun et al. [43], an elastic perfectly plastic material model with Von Mises yield criterion σ_y (GPa) is assumed for the indented materials. The yield stresses are considered to be equal to 0.198 GPa for LD C—S—H, 0.347 GPa for HD C—S—H, and 0.440 GPa for CH.

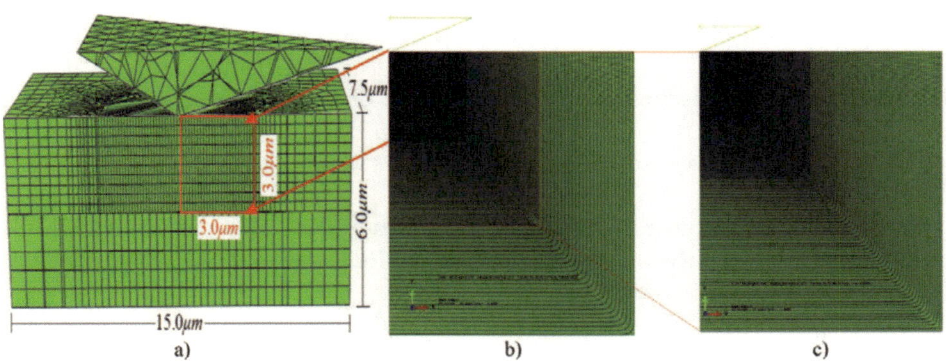

Figure 7.
Indentation simulation on 3-D and 2-D models of typical cement paste phases. (a) 3-D geometrical model. (b) 2-D model with three mesh regions. (c) Partial 2-D model with fine mesh.

Phases/parameters	LD C—S—H	HD C—S—H	CH
E (GPa)	17.35	32.70	39.88
Friction coefficient *f*	0.6	0.6	0.6
μ	0.2–0.3	0.2–0.3	0.2–0.3

Table 1.
Mechanical properties and parameters of typical phases and structures used in simulation: parameters of the representative experimental CH and C—S—H phases.

Phases/parameters	LD C—S—H		HD C—S—H		CH	
	S-C scheme	M-T scheme	S-C scheme	M-T scheme	AFEM	DFT + RVH
Elastic modulus *E* (GPa)	18.22	29.78	31.63	37.71	43.13	45.46
Poisson's ratio μ	0.25	0.25	0.25	0.25	0.25	0.20
Friction coefficient *f*	0.6	0.6	0.6	0.6	0.6	0.6

Table 2.
Mechanical properties and parameters of typical phases and structures used in simulation: parameters of the CH, LD C—S—H, and HD C—S—H structures using various simulation techniques.

Frictional effects in the indenter-material interface were included in the analysis through an isotropic model. We assume that the loading rate is slow enough such as static friction can securely model the interface response. Simulations are proceeded in two steps. The indenter was first subjected to a ramped vertical displacement, followed by an indenter retraction to the original position which corresponded to complete unloading at zero load. Moreover, only half of the material is chosen to simulate nanoindentation with fraction coefficients between indenter and material adopted. Axisymmetric boundary conditions were used along the symmetry axis beneath the indenter. During simulation, five degrees of freedom are limited except *y*-axis and the bottom plane (constrained vertically) and planes around are fixed, then material model is defined. The contact mode is used during loading and unloading processes.

For comparative analysis, the FEM simulation of nanoindentation on the LD C—S—H, HD C—S—H, and CH structures using parameters in **Tables 1** and **2** are carried out using ABAQUS.

5. Discussion on elastic moduli by both experiment and simulation

5.1 Experiment and discussion of elastic moduli of CH and C—S—H

5.1.1 Nanoindentation experiment and regional indentation analysis

After the sample preparation and chemical composition, nanoindentation experiment is done on the instrument of INSA de Rennes, and CSM instrument and sample are shown in **Figure 8**.

As in **Figure 8(a)**, CSM instruments include: ultrananoindentation tester head, nanoindentation tester head, atomic force microscopy, and optical video microscope. Besides, Berkovich indenter is shown in **Figure 8(b)**. Samples with the dimension 1.1 × 1.1 × 1.1 cm underwent mechanical polishing and electrolytic polishing separately and lapped to a mirror, as shown in **Figure 8(c)**. Parameters

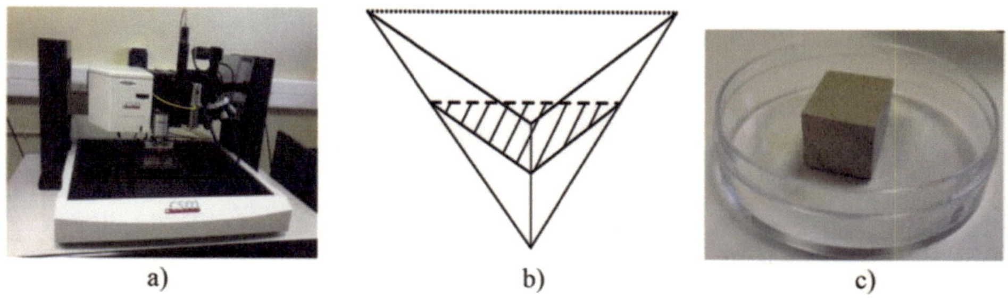

Figure 8.
CSM instrument and cement paste used in experiment. (a) Nanohardness platform. (b) Berkovich indenter. (c) Cement paste sample.

are as: the maximum load is 1.5 mN; velocity after contact with the sample surface is 10 m/s; continuous maximum load is 5 s; unloading time is 10 s. The speed of loading/unloading was 9.6841 mN/s and the *P-h* data were recorded. The test is repeated several times to average.

5.1.2 Elastic moduli of typical phases in various samples

Nanoindentation experiments were carried out on CSM microindentation tester with conical Berkovich indenter. The process is mainly described as follows: samples with the dimensions 1.1 × 1.1 × 1.1 cm underwent mechanical polishing and electrolytic polishing separately and lapped to a mirror. Then, experimental maximum loads of 1.5 and 2 mN were mainly selected by loading/unloading with the speed of 9.6841 mN/s and the data of load and depth were recorded. Besides, indentation diagonal lengths were measured by data collection software on computer. Under each parameter, experiments were repeatedly done five times and then averaged to decrease the relative error. By experimentally investigating the mechanical properties of cement paste at different length scales provides a means of correlating such nanoscale properties to nanoscale applications. The identification of the parameters is provided in **Table 3**.

For the tested curves of CH phase collected from different samples, the parameters of CH phase obtained microindentation experiment are shown in **Table 4**.

The representative experimental curves of typical phases characterized by individual elastic modulus are averaged. Using the indents of LD C—S—H phase under different samples, the corresponding *P-h* curves are found. The representative experimental curves of outer LD C—S—H phase are given in **Figure 9**.

Experimental parameters	Unit	CEMI	CEMIII	CEMI_MK	CEMI_Sed
Maximum load P_{max}	mN	1.5–2.0	1.5	1.5	1.5–2.0
Strain rate P^{*}/P_{max}	s^{-1}	0.01	0.01	0.01	0.01—
Loading and unloading speed	mN/min	3–4	3–4	2–5	2–5
Acquisition frequency rate	Hz	10	10	10	10
Delta slope (%)	—	80	80	80	80
Poisson's ratio	—	0.2	0.2	0.2	0.2
Total number of indents	—	830	363	334	296

Table 3.
Characteristics of experimental parameters used in loading programs for samples.

Experimental parameters	LD C—S—H	HD C—S—H	CH	Calcite	Clinker
Maximum depth h_{max} (nm)	441.79	278.08	212.98	97.11	99.54
The projected contact area A_p (nm^2)	4577259.84	1761371.10	977898.05	137943.39	267721.30
The unloading stiffness S (mN/nm)	0.0532	0.0559	0.0458	0.0363	0.0588
The fitting parameter m	4.76	6.01	4.12	1.29	1.66
Indent number (e.g., CEMIII, etc.)	30	22	56	13	—

Table 4.
Experimental parameters and corresponding characteristics of typical phases.

Figure 9 shows the representative experimental curve of LD C—S—H phase, with elastic modulus of 17.35 GPa. Similarly, representative experimental curves of HD C—S—H phase is shown in **Figure 10**.

Figure 10 shows the representative experimental curve of HD C—S—H phase, with elastic modulus of 32.70 GPa. Similarly, the representative corresponding *P-h* curve of CH phases is given in **Figure 11**.

Figure 11(b) shows the representative experimental curve of CH phase, with its corresponding elastic modulus of 39.88 GPa.

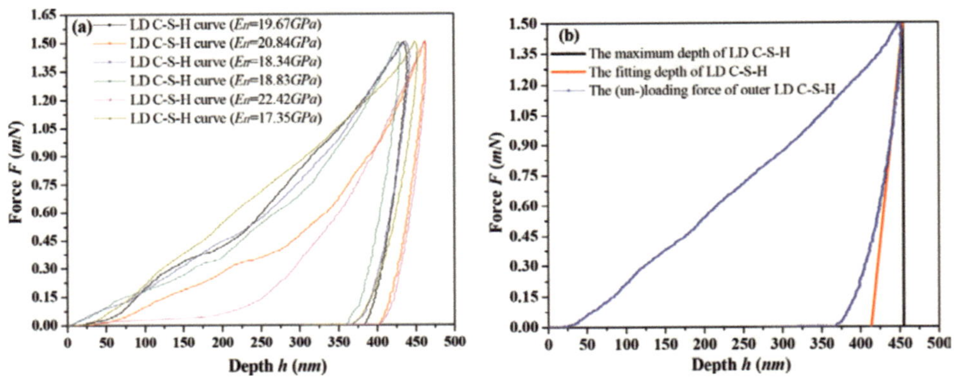

Figure 9.
The representative experimental curves of outer LD C—S—H phase. (a) Typical curves of outer LD C—S—H phase. (b) The representative curve of outer LD C—S—H phase.

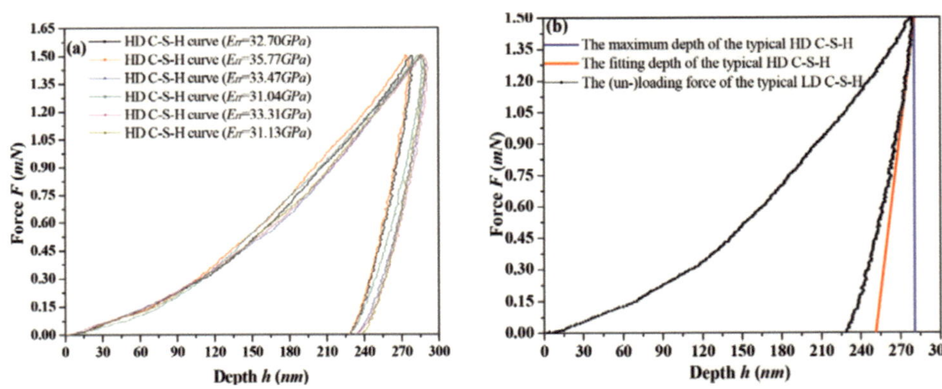

Figure 10.
The representative experimental curves of inner HD C—S—H phase. (a) Typical curves of inner HD C—S—H phase. (b) The representative curve of inner HD C—S—H phase.

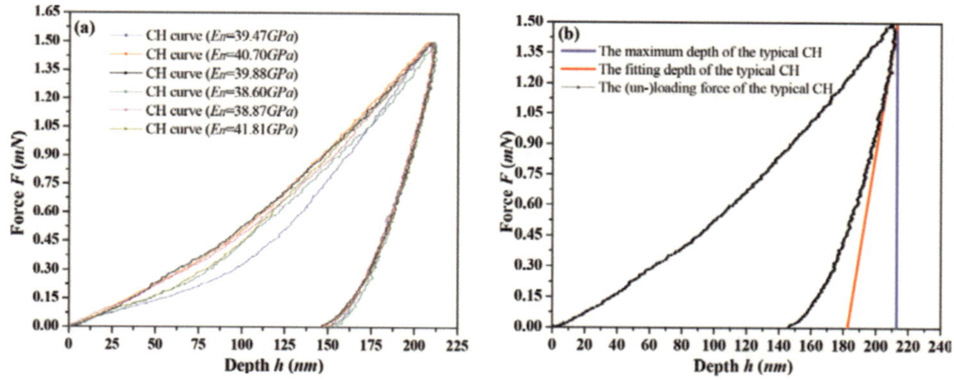

Figure 11.
The representative experimental curves of CH phase. (a) Typical curves of CH phase. (b) The representative curve of CH phase.

Overall, nanoindentation experiments show that the range of elastic modulus corresponding to low-density C—S—H is 15–26 GPa and the range of elastic moduli corresponding to high-density C—S—H is 26–39 GPa. The inverse method described above is applied to each matrix phase that is what has been called previously: LD C—S—H, HD C—S—H, portlandite (CH), and clinker.

5.2 Simulation and discussion of elastic moduli of CH and C—S—H

There are two methods of applying load in simulation: forced loading and forced displacement. Comparison of the load-displacement curves using the forced loading and the forced displacement is the same; here, we use the forced displacement method.

5.2.1 Simulation and discussion of elastic modulus of CH phase

From the load-depth curve of CH phase in nanoindentation experiment, the pressure is about 1.5 mN and the maximum indent depth of CH phase is about 212.98 nm, the representative curve of CH is as in Ref. [44]. CH simulation parameters were separately set as: elastic moduli E are separately 45.459 GPa (DFT + RVH method, μ = 0.2), 43.13 GPa (AFEM method, μ = 0.25), and 39.880 GPa (nanoindentation experiment). The region range of Poisson ratio μ is 0.25–0.30. The yield stress can be averaged by the experimental data [21] and the simulated data [38], which is 0.440 GPa by the Eq. (27). The contact friction is set about 0.6 [26].

Stress distribution and deformed displacement zones of CH are shown in **Figure 12**.

As is shown in **Figure 12**, the maximum residual stress after unloading on the ideal contact condition distributes just below the indenter, while the maximum residual stress after unloading on the friction condition distributes around the top of indenter and the residual stress below the indenter is not too large.

Using ABAQUS software, under the loading force of 1.5 mN, the comparison of experimental and simulated unloading *P-h* curves of CH phase is shown in **Figure 13**.

As is shown in **Figure 13**, we can see that the simulated unloading part is close to the representative experimental part. It can be seen that the smaller Poisson's ratio is, the smaller the slope of unloading curve will be. Comparison of purely elastic unloading curve of CH phase is shown in **Figure 13**; we can see that DFT methods (45.459 GPa, by LDA method) and AFEM method (43.13 GPa) are very consistent

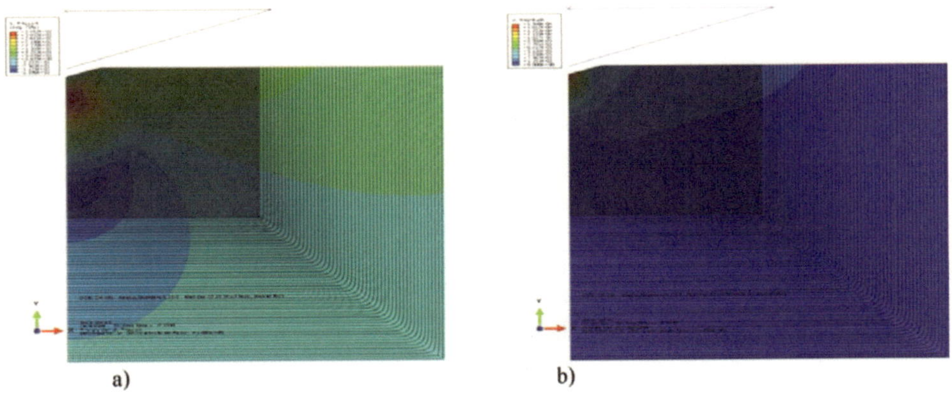

Figure 12.
Stress distribution and displacement zones of CH under various frictions. (a) Stress distribution zone (f = 0.6). (b) Deformed displacement zone (f = 0.6).

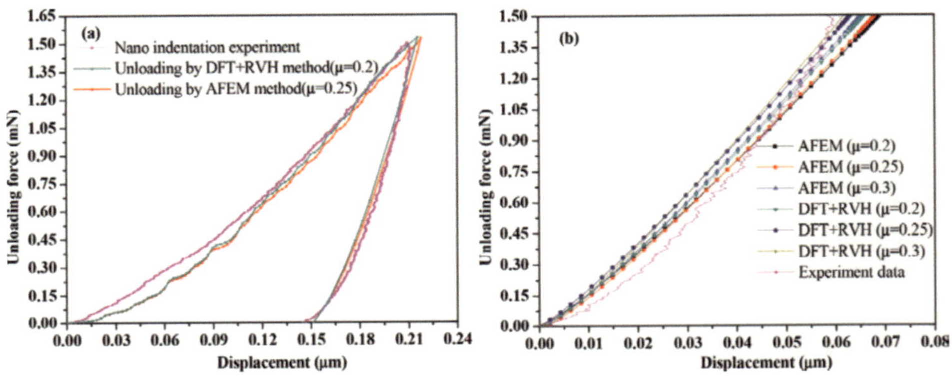

Figure 13.
Comparison of experimental and simulated P-h and unloading curves of CH phase. (a) Comparison of experiment and AFEM and DFT methods. (b) Unloading curves under various Poisson's ratio.

with experimental results (39.880 GPa), which verifies the reliability of simulation method.

For CH structure in this book, the elastic constants have been calculated using the AFEM and DFT methods. Then, the homogenized elastic modulus of CH structure is estimated by the Y parameter, elastic moduli by DFT method (45.459 GPa) and by AFEM (43.14 GPa) are very close to the results of Constantinides. In order to verify the simulation results of the CH structure, nanoindentation experiments of four cement samples are carried out. The averaged Young's modulus of CH phase is 39.880 GPa by nanoindentation experiment, which is close to the results of 40.3 $\pm$ 4 GPa by Constantinides. Comparison of purely elastic unloading curve of CH phase shows that DFT and AFEM methods are very consistent with experimental results on average, which verifies the reliability of both simulation methods. It has demonstrated the reasonableness of simulation conclusions by the reliable experiment verification.

The simulated elastic modulus is close to the average by experiment, which has demonstrated the reasonableness of simulation conclusions. Although the mechanism of the complex plastic deformation is still unknown, the yield flow stress parameter is still turned to be of certain rationality. The nanoindentation simulation of CH phase is done on ABAQUS using the corresponding parameters, thus reversely verify the reliability of the initial parameters. The large-scale models using

QC methods are considered and verified on ABAQUS software, where, on behalf of the purely elastic stage, the unloading curves of the CH model with various Poisson's ratios are simulated. It can be seen that the smaller Poisson's ratio is, the smaller the linear slope of unloading curve will be.

5.2.2 Simulation and discussion of elastic modulus of LD C—S—H

Firstly, elastic constants of the monoclinic 11 Å tobermorite are calculated by DFT. Then, the homogenized elastic modulus is determined to be 79.51 GPa, which is close to the other references. However, the ordered 11 Å tobermorite model cannot reflect the C—S—H structure, and the crystalline C—S—H structures—C—S—H(I) (Ca/Si = 0.67) and the C—S—H(II) (Ca/Si = 1.67)—are still far from the data in laboratory. That is why, the disordered structure is simulated and confirmed. For the monolithic structures of C—S—H(I) and C—S—H(II), using MD simulation, the averaged Young's moduli are separately about 75.92 and 60.95 GPa. The last value of C—S—H(II) with the size about 5 nm enables to evaluate the value of the elastic modulus of LD C—S—H and HD C—S—H: 18.11 and 31.45 GPa, using the self-consistent scheme.

The LD C—S—H phase can be modeled assuming a linear elastic and cohesive-frictional plastic material. The forced displacement method is used as mentioned above.

From the load-depth curve of LD C—S—H phase nanoindentation experiment, the pressure is about 1.5 mN and the maximum indent depth is about 441.79 nm, the representative curve of LD C—S—H is in Ref. [44]. The yield stress can be averaged by the experimental data [21] and the simulation [38], which is 0.198 GPa. The LD C—S—H parameters were separately set as follows: elastic moduli E are separately 18.22 GPa (by the self-consistent scheme, μ = 0.25), 29.78 GPa (by the Mori-Tanaka scheme, μ = 0.25), and 17.35 GPa (nanoindentation experiment). The region range of Poisson ratio μ is 0.20–0.25 and the contact friction is about 0.6 [26].

Stress distribution and deformed displacement zones of LD C—S—H are shown in **Figure 14**.

As is shown in **Figure 14**, the maximum residual stress after unloading on the ideal contact condition distributes just below the indenter, while the maximum residual stress after unloading on the friction condition distributes around the top of indenter and the residual stress below the indenter is not too large. Similarly, the

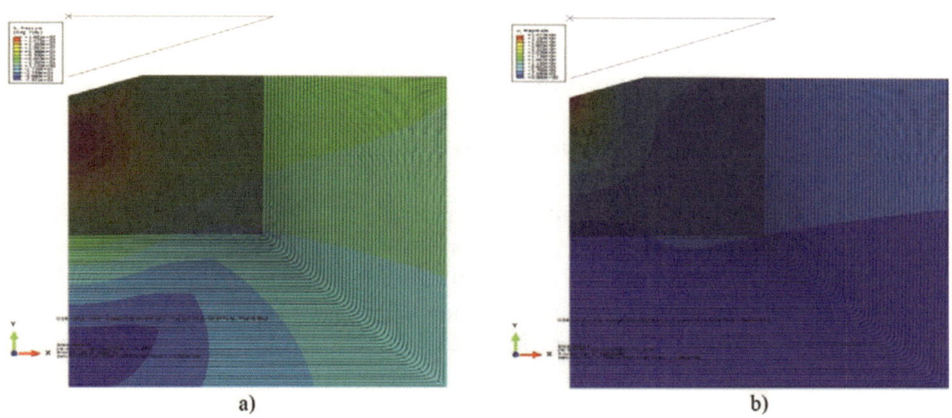

Figure 14.
Stress distribution and displacement zones of LD C—S—H under various frictions. (a) Stress distribution zone (f = 0.6). (b) Deformed displacement zone (f = 0.6).

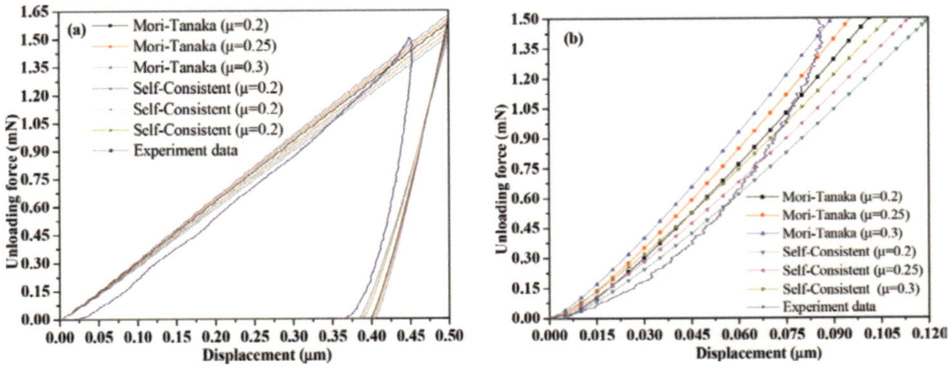

Figure 15.
Comparison of experimental and simulated unloading P-h curves of LD C—S—H phase. (a) Comparison of experiment and MD simulation. (b) Unloading curves under various Poisson's ratio.

comparison of experimental and simulated unloading *P-h* curves of LD C—S—H phase is shown in **Figure 15**.

As is shown in **Figure 15(a)**, we can see that the simulated unloading part is close to the representative experimental part. It can be seen that the smaller Poisson's ratio is, the smaller the slope of unloading curve will be. Comparison of purely elastic unloading curve of LD C—S—H phase is shown in **Figure 15(b)**; we can see that, the Mori-Tanaka scheme (29.78 GPa, μ = 0.25) and the self-consistent scheme (18.22 GPa, μ = 0.25) are very consistent with experimental results (17.35 GPa), which verifies the reliability of both simulation methods. Comparison of purely elastic unloading curve of LD C—S—H phase shows that the simulated *P-h* curve (18.22 GPa, by the self-consistent scheme of C—S—H(II) structure (C/S = 1.67) with the size about 5 nm simulated using MD method in Chapter 5) is close to the experimental curve (17.35 GPa), which verifies the reliability of both simulation methods.

5.2.3 Simulation and discussion of elastic modulus of HD C—S—H

From the representative load-depth curve of HD C—S—H phase nanoindentation experiment, the pressure is about 1.5 mN and the maximum indent depth of HD C—S—H phase is about 278.08 nm, the representative curve of HD C—S—H is in Ref. [44]. Similarly, the HD C—S—H parameters were separately set as follows: elastic moduli E are separately 31.63 GPa (by the self-consistent scheme, μ = 0.25), 37.71 GPa (by the Mori-Tanaka scheme, μ = 0.25), and 32.70 GPa (nanoindentation experiment). The region range of Poisson ratio μ is 0.20–0.25, and the yield stress in the Eq. (27) is 0.347 GPa [38]. By the Ref. [26], the contact friction is about 0.6.

Stress distribution and deformed displacement zones of HD C—S—H are in **Figure 16**.

As is shown in **Figure 16**, the maximum residual stress after unloading on the ideal contact condition distributes just below the indenter, while the maximum residual stress after unloading on the friction condition distributes around the top of indenter and the residual stress below the indenter is not too large.

Using ABAQUS software, under the loading force of 1.5 mN, the comparison of experimental and simulated unloading *P-h* curves of HD C—S—H phase is shown in **Figure 17**.

As is shown in **Figure 17(a)**, we can see that the simulated unloading part is close to the representative experimental part. It can be seen that the smaller Poisson's ratio is, the smaller the slope of unloading curve will be. Comparison of

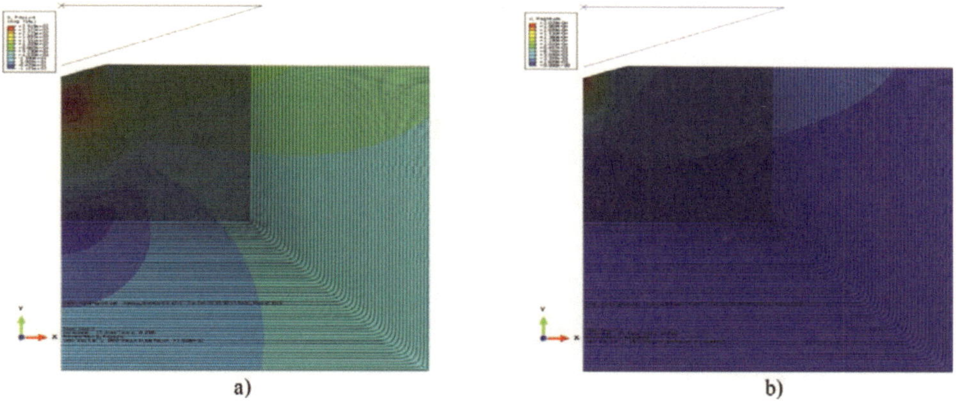

Figure 16.
Stress distribution and displacement zones of HD C—S—H under various frictions. (a) Stress distribution zone (f = 0.6). (b) Deformed displacement zone (f = 0.6).

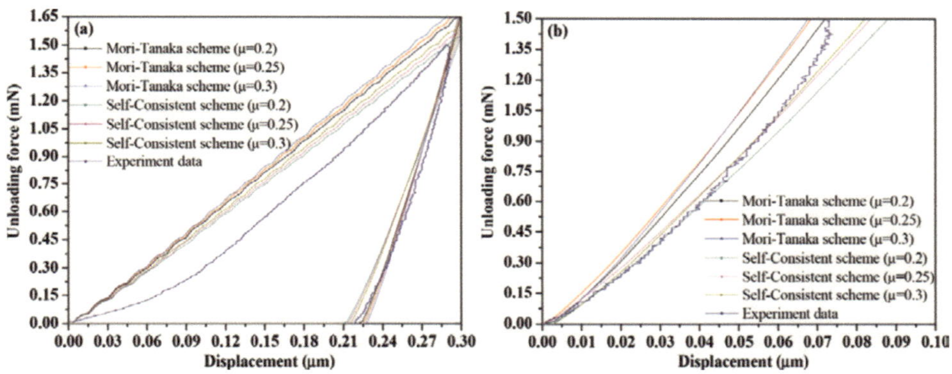

Figure 17.
Comparison of experimental and simulated unloading P-h curves of HD C—S—H phase. (a) Comparison of experiment and MD simulation. (b) Unloading curves under various Poisson's ratio.

purely elastic unloading curve of HD C—S—H phase is shown in **Figure 17(b)**; we can see that the Mori-Tanaka scheme (37.71 GPa, μ = 0.25) and the self-consistent scheme (31.36 GPa, μ = 0.25) are very consistent with experimental results (32.70 GPa), which verifies the reliability of simulation method. This indicates that the selected simulation parameters are reliable to characterize its plastic deformation behaviors. It is found that the comparison results of the experimental and simulated unloading curves are basically coincided and have a quite good agreement. Thus, the comparison of the experimental results and simulated values shows the feasibility and rationality of the simulation methods mentioned above.

In order to verify the simulation results of the structures, especially for the LD C—S—H, nanoindentation simulation is carried out. The simulated elastic modulus is close to the average by experiment, which has demonstrated the reasonableness of simulation. From the simulation results and experimental data [44], it indicates that the selected simulation parameters are reliable, which thus provides the basis and parameters for the multiscale simulation [35, 37].

In conclusion, the properties of each phase contained in the cement paste are considered in such a way that hydrated and nonhydrated cement phases are assumed to be perfectly elastic. The values of the elastic modulus and of the Poisson coefficient of these different phases are obtained from nanoindentation

Phases	Density (g/cm^3)	Mixing volume (%) [45]	Paste volume (%)[45]	*E* (GPa) by reference	*E* (GPa) by simulation	*E* (GPa) by experiment
C_3S	3.15	23.40	1.17	117.6 [47]	—	—
C_2S	3.28	7.35	0.78	117.6 [47]	—	—
C_3A	3.03	4.42	0.00	117.6 [47]	—	—
C_4AF	3.73	2.87	1.39	117.6 [47]	—	—
Gypsum	2.32	3.47	0.00	45.7 [48]	—	—
Portlandite (CH)	2.24	0.00	13.96	46.6 [49]	43.13 (AFEM); 45.46 (DFT)	≈44.7
C—S—H	1.90	0.00	29.03	22.4 [50]	Monolithic C—S—H: C—S—H(I) ≈ 75.92; C—S—H(II) ≈ 60.95	≈22.05
C—S—H pozz	2.65	0.00	49.99	22.4 [50]	By SC scheme: LD C—S—H = 18.22; HD C—S—H = 31.63	≈33.52
Afm	2.02	0.00	15.12	42.3 [50]	—	—
Ettringite (AFt)	1.78	0.00	6.87	22.4 [50]	—	—
Porosity total	0	58.49	31.69	0	—	—

Table 5.
Young's modulus of phases in Portland cement paste by references and our work.

measurements. The comparison of elastic modulus between references [45, 46] and present work in thesis is shown in **Table 5**.

The nanoindentation technique has identified in various phases of the cement pates to determine their elastic moduli. The unloading *P-h* curves under various Poisson's ratio have been finally obtained, which provides basis for theoretical guidance and practical analysis.

6. Conclusions

Nanoindentation simulation is done on ABQUS to get mechanical properties of CH and C—S—H, with the comparison of nanoindentation experiment. The effect of Poisson's ratio on the nanoindentation simulation of CH and C—S—H has been quantified through FE simulations. A numerical model was developed to validate and assess the elastic parameters associated with each phase. The *P-h* curves of the simulation are compared to conclude on the feasibility of the methodology. The experimental elastic moduli of LD C—S—H and HD C—S—H deviate up to ±20% in our previous work. The present simulation helps to understand more advanced constitutive relations for C—S—H (time-dependent deformation, yield-criteria, hardening phenomena, etc.).

1. The simulated Young's modulus of CH phase is about 45.459 GPa (DFT methods, by LDA method) and 43.13 GPa (AFEM method) by nanoindentation simulation, which is close to the results of 40.3 ± 4 GPa by Constantinides.

2. Based on the evaluated elastic modulus values of LD C—S—H 18.11 GPa and HD C—S—H 31.45 GPa using the self-consistent scheme of the monolithic C—S—H(II) structure 60.95 GPa with the size about 5 nm, the simulated results for LD C—S—H structure 18.22 GPa and HD C—S—H structures 31.63 GPa, which is close to the results of 18.2 $\pm$ 4.2 and 29.4 $\pm$ 2.4 GPa by Constantinides, thus has demonstrated the reasonableness of simulation parameters.

3. Finally, on behalf of the purely elastic stage, the unloading curves of the three (CH, LD C—S—H, and HD C—S—H structures) models are simulated on ABAQUS software. It is found that the comparison results of the experimental and simulated unloading curves are basically coincided and have a quite good agreement. Thus, the analysis of unloading curves shows the feasibility and rationality of the simulation methods mentioned above.

Overall, as the length scale of C—S—H structure with hundreds of nanometers can be called C—S—H gel, and these colloids have a number of C—S—H particles having an average particle radius about $\sim$5 nm, the C—S—H model composed of C—S—H particles with radius of 5 nm for nanoindentation simulation thus can also be precisely described by MD in future, which can be used to compare the results using FEM method, although simulation of these models at larger scale is time-consuming. In the future, the upscaling of the obtained properties at the atomic level will be concerned in order to connect this modeling to continuum models by using relevant input form at small level, carrying forward the critical information to represent the continuum model with the intrinsic nanoscale features incorporated. Other mechanical properties may also be investigated, not only fracture properties and their upscaling toward upper scales, but also creep behavior of the C—S—H phase which exhibits such a time-dependent behavior.

Acknowledgements

The author acknowledges the financial support provided by the China Scholarship Council and the startup foundation of Xi'an Shiyou University. Thanks Qiufeng Wang for her proofreading, and very sincere thankness to prof. Fabrice Bernard and prof. Siham Kamali-Bernard for the their enthusiastic and responsible guideness.

Author details

Jia Fu[1,2]

1 Xi'an Shiyou University, Xi'an, China

2 INSA de Rennes, Rennes, France

*Address all correspondence to: fujia@xsyu.edu.cn

References

[1] Liu WK, Karpov EG, Park H. Nano Mechanics and Materials: Theory, Multiscale Analysis and Applications. New York: Wiley; 2005

[2] Maekawa K, Ishida T, Kishi T. Multi-scale modeling of concrete performance. Journal of Advanced Concrete Technology. 2003;**1**(2):91-126

[3] Curtin WA, Miller RE. Atomistic/continuum coupling in computational materials science. Modelling and Simulation in Materials Science and Engineering. 2003;**11**:33-68

[4] Fish J. Bridging the scales in nano engineering and science. Journal of Nanoparticle Research. 2006;**8**(5):577-594

[5] Lu G, Kaxiras E. In: Rieth M, Schommers W, editors. Overview of Multiscale Simulations of Materials. Handbook of Theoretical and Computational Nanothechnology. Vol. X. California Valencia: American Scientific Publishers; 2005. pp. 1-33

[6] Gilman JJ. Chemistry and Physics of Mechanical Hardness. Hoboken: John Wiley & Sons; 2009

[7] Oliver WC, Pharr GM. Measurement of hardness and elastic modulus by instrumented indentation: Advances in understanding and refinements to methodology. Journal of Materials Research. 2004;**19**(01):3-20

[8] Riedel R. Handbook of Ceramic Hard Materials. Weinheim: Wiley-VCH; 2000

[9] Geiger M, Kleiner M, Eckstein R. Microforming. CIRP Annals - Manufacturing Technology. 2001;**50**(2): 445-462

[10] Al-Rub RKA, Voyiadjis GZ. Analytical and experimental determination of the material intrinsic length scale of strain gradient plasticity theory from micro- and nano-indentation experiments. International Journal of Plasticity. 2004;**20**:1139-1182

[11] Tho KK et al. Numerical simulation of indentation with size effect. Materials Science and Engineering A. 2006;**421** (1–2):268-275

[12] Fleck NA et al. Strain gradient plasticity: Theory and experiment. Acta Metallurgica et Materialia. 1994;**42**(2): 475-487

[13] Huber N, Nix WD, Gao H. Identification of elastic-plsatic material parameters from pyramidal identation of thin films. Proceedings of the Royal Society of London A. 2002;**458**(2023): 1593-1620

[14] Gao H, Huang Y, Nix WD, Hutchinson JW. Mechanism-based strain gradient plasticity-I. Theory. Journal of the Mechanics and Physics of Solids. 1999;**47**:1239-1263

[15] Zhijian P, Jianhong G, Hezhan M. On the description of indentation size effect in hardness testing for ceramics: Analysis of the nanoinfentation data. Journal of the European Ceramic Society. 2004;**24**:2193-2201

[16] Han L, Hu H, Northwood DO. Effect of Ca additions on microstructure and microhardness of an as-cast Mg-5.0 wt.% Al alloy. Materials Letters. 2008; **62**(3):381-384

[17] Arnaud Charles Albert Muller. Characterization of porosity & CSH in cement pastes by 1H NMR [Thèse de doctorat]. École Polytechnique Fédérale de Lausanne; 2014

[18] Ye G, Van Breugel K, Fraaij ALA. Three-dimensional microstructure analysis of numerically simulated cementitious materials. Cement and Concrete Research. 2003;**33**(2):215-222

[19] Ulm F-J, Vandamme M, Bobko CP, Ortega JA, Tai K, Qrtiz C. Statistical indentation techniques for hydrated nanocomposites: Concrete, bone, and shale. Journal of the American Ceramic Society. 2007;**90**(9):2677-2692

[20] Sorelli L, Constantinides G, Ulm F-J, et al. The nano-mechanical signature of ultra-high performance concrete by statistical nanoindentation. Cement and Concrete Research. 2008;**38**:1447-1456

[21] Vandamme M. The nanogranular origin of concrete creep: A nanoindentation investigation of microstructure and fundamental properties of calcium-silicate-hydrates [PhD thesis]. Cambridge, MA: Massachusetts Institute of Technology; 2008

[22] Zanjani Zadeh V, Bobko CP. Nano-mechanical properties of internally cured kenaf fiber reinforced concrete using nanoindentation. Cement and Concrete Composites. 2014;**52**:9-17

[23] Vandamme M, Ulm FJ. Nanoindentation investigation of creep properties of calcium silicate hydrates. Cement and Concrete Research. 2013; **52**:38-52

[24] Oliver WC, Pharr GM. An improved technique for determining hardness and elastic modulus using load and displacement sensing indentation experiments. Journal of Materials Research. 1992;7(06):1564-1583

[25] Keinde D. Etude du béton à l'échelle mesoscopique: Simulation numérique et tests de micro-indentation [Doctoral dissertation]. INSA de Rennes; 2014

[26] Constantinides G, Ulm FJ. The effect of two types of CSH on the elasticity of cement-based materials: Results from nanoindentation and micromechanical modeling. Cement and Concrete Research. 2004;**34**(1): 67-80

[27] Sneddon IN. The relation between load and penetration in the axisymmetric boussinesq problem for a punch of arbitrary profile. International Journal of Engineering Science. 1965; **3**(1):47-57

[28] Guillonneau G. Nouvelles techniques de nano-indentation pour des conditions expérimentales difficiles: Très faibles enfoncements, surfaces rugueuses, température [Doctoral dissertation]. Ecole Centrale de Lyon; 2012

[29] Huang Y, Zhang F, Hw ang KC, et al. A model of size effects in nano-indentation. Journal of the Mechanics and Physics of Solids. 2006;**54**(8): 1668-1686

[30] Nix WD, Gao H. Indentation size effects in crystalline materials: A law for strain gradient plasticity. Journal of the Mechanics and Physics of Solids. 1998; **46**(3):411-425

[31] Dao M, Chollacoop N, Van Vliet KJ, et al. Computational modeling of the forward and reverse problems in instrumented sharp indentation. Acta Materialia. 2001;**49**:3899-3918

[32] Qu S, Huang Y, Pharr GM, Hwang KC. The indentation size effect in the spherical indentation of iridium: A study via the conventional theory of mechanism-based strain gradient plasticity. International Journal of Plasticity. 2006;**22**(7):1265-1286

[33] Tabor D. A simple theory of static and dynamic hardness. Proceedings of the Royal Society A: Mathematical, Physical and Engineering Sciences. 1948:247-274

[34] Bernard F, Kamali-Bernard S, Prince W. 3D multi-scale modelling of mechanical behavior of sound and leached mortar. Cement and Concrete Research. 2008;**38**:449-458

[35] Workshop of Multiscale Modelling Course for Concrete MMC, TU Delft

[36] Kamali-Bernard S, Bernard F. Effect of tensile cracking on diffusivity of mortar: 3D numerical modelling. Computational Materials Science. 2009; **47**:178-185

[37] Sab K. On the homogenization and the simulation of random materials. European Journal of Mechanics - A/ Solids. 1992;**11**(5):585-607

[38] Sarris E, Constantinides G. Finite element modeling of nanoindentation on C—S—H: Effect of pile-up and contact friction. Cement and Concrete Composites. 2013;**36**:78-84

[39] Pettermann H. Continuum Mechanics Modeling of Composite materials. TU-Wien: Habilitationsschrift; 2003

[40] Heqing L, Xun C, Qiulong C. Characterization of mechanical properties of thin films using nanoindentation technique. Transactions of Materials and Heat Treatment. 2001;**22**(4):52-55

[41] Lu YC, Shinozaki DM. Effects of substrate constraint on micro-indentation testing of polymer coatings. Materials Science and Engineering A. 2005;**396**(1–2):77-86

[42] Shi Z et al. The equivalent axisymmetric model for Berkovich indenters in power-law hardening materials. International Journal of Plasticity. 2010;**26**(1):141-148

[43] Asroun N, Asroun A. Nanoindentation simulation of concrete with various indenter forms and yield strengths. International Journal of Civil and Structural Engineering. 2013;**1**(4): 36-45

[44] Fu J, Kamali-Bernard S, Bernard F, Cornen M. Comparison of mechanical properties of C—S—H and Portlandite between nano-indentation experiments and a modelling approach using various simulation techniques. Composites Part B: Engineering. 2018;**151**:127-138

[45] Kwan S, LaRosa Thompson J, Grutzeck MW. Structures and phase relations of aluminum-substituted calcium silicate hydrate. Journal of the American Ceramic Society. 1996;**79**(4): 967-971

[46] Kamali S, Moranville M, Garboczi EG, et al. Hydrate dissolution influence on the young's modulus of cement paste. In: 5th International Conference of Fracture Mechanics of Concrete Structures, Vail, CO. 2004. pp. 12-16

[47] Boumiz A, Sorrentino D, Vernet C, et al. Modelling the development of the elastic moduli as a function of the hydration degree of cement pastes and mortars. In: Proceedings of 2nd International RILEM Symposium, Hydration and Setting, RILEM. France: Cachan Cedex; 2000. pp. 295-316

[48] Choy MM, Hellwege KH, Hellwege AM. Elastic, Piezoelectric, Pyroelectric, Piezooptic, Electrooptic Constants, and Nonlinear Dielectric Susceptibilities of Crystals: Revised and Expanded Edition of Volumes III/1 and III/2. Berlin Heidelberg: Springer-Verlag; 1979

[49] Holuj F, Drozdowski M, Czajkowski M. Brillouin spectrum of $Ca(OH)_2$. Solid State Communications. 1985;**56**(12): 1019-1021

[50] Damidot D, Velez K, Sorrentino F. Characterisation of interstitial transitionzone (ITZ) of high performance cement by nanoindentation technique. In: 11th International Congress on the Chemistry of Cement, Durban. 2003